PHYSICS

IN

100

NUMBERS

METRO BOOKS
New York

An Imprint of Sterling Publishing
1166 Avenue of the Americas
New York, NY 10036

Conceived, designed, and produced by
Quid Publishing
Level 4 Sheridan House
114 Western Road
Hove BN3 1DD
England

Design and illustration by Simon Daley

Dedication: For Ruth, the best collection of atoms in the
universe

www.quidpublishing.com

ISBN: 978-1-4351-5801-6

Manufactured in China

1 3 5 7 9 10 8 6 4 2

www.sterlingpublishing.com

PHYSICS

IN

100

NUMBERS

A Numerical Guide to Facts,
Formulas, and Theories

COLIN STUART

METRO BOOKS

Table of contents

Introduction

A note on numbers...

As a subject, physics tries to grapple with some of the biggest questions in the universe. That could be the inner workings of the atom or the motion of huge clusters of galaxies. Such a span of distance scales means that the numbers involved can often be very small or dizzyingly large.

Physicists (and other scientists and mathematicians) have a shortcut for dealing with these otherwise unmanageable numbers: standard form.

Take the number 0.00000000000000000000737, for example. Writing it as 7.37×10^{-20} makes life much easier. Similarly, 7,370,000,000,000,000,000,000 is much more neatly written as 7.37×10^{20}. This convention will be used extensively in the pages that follow.

A note on units...

The official system of units in physics is known as the International System of Units or SI system (from the French *Le Système International d'Unités*). There are only seven base units from which all others can be derived. They are: the meter, the kilogram, the second, the ampere, the kelvin, the mole, and the candela.

Such units have been used throughout where convenient to do so. Sometimes it pays not to write everything fully in SI units.

Take the permittivity of free space, for example (see page 26) which has stated units of coulombs squared per newton meter squared (C^2/Nm^2). Coulombs and newtons are not base units. Converting everything into base units would give $A^2s^4/kg\ m^3$ which is messier.

A note on orders of magnitude...

Units can have prefixes which denote fractions or multiples of that unit. Common everyday examples are the millimeter (one thousandth of a meter) and the kilometer (one thousand meters). Here's a list of the ones used most often.

giga	G	10^9
mega	M	10^6
kilo	k	10^3
centi	c	10^{-2}
milli	m	10^{-3}
micro	μ	10^{-6}
nano	n	10^{-9}

5.39×10^{-44}

Planck time (s)

In the modern world, timing is everything. Racing drivers and athletes sweep to glory by a margin that's often only hundredths of a second. In 2010, a company spent millions of dollars laying improved fiber-optic cable between Chicago and New York, just to shave three milliseconds off the time it takes for financial information to ping between the cities. As we'll see, the satellites of the Global Positioning System deal in time signals accurate to millionths of a second.

You cannot keep carving up time, however. The smallest meaningful time interval is known as the Planck time. It forms part of a system of physical units put forward by the German physicist Max Planck in 1899. He was keen to define a set of natural units, based purely on fundamental constants of nature, rather than human experience. Conventional, everyday units of time—such as days, months, and years—are based on the motions of the Earth, Sun, and Moon.

One unit of Planck time marks the earliest point in the universe's history at which our current laws of physics make any sense. Try to probe earlier than this and the rules that govern the geometry of space—Einstein's Theory of General Relativity—break down. So our current picture of the early universe cannot begin at time zero (t=0). We're restricted instead to starting our biographical account from t=5.39×10^{-44} seconds.

1.62×10^{-35}

Planck length (m)

Similar to the Planck time, the Planck length is the smallest distance over which the current laws of physics still hold. Beyond this point, space descends into a fuzzy haze that physicists refer to as the "quantum foam." It is also the distance covered by a beam of light in a time period equal to the Planck time.

Like all units in Planck's system, it is calculated by dealing only with fundamental constants of nature—in this case, Isaac Newton's Universal Gravitational Constant, G (see page 28), Planck's own constant (see page 12) and the speed of light (see page 126).

▲ Introducing an indivisible distance such as the Planck length is one solution to Zeno's paradox, which suggests runners can never finish races.

The never-ending race

Consider this ancient riddle known as "Zeno's paradox." Imagine a group of sprinters lined up for the 100-meter Olympic final. To finish the race, they first need to run halfway. Once they reach this 50-meter (164-foot) mark, they must then run half the remaining distance. Having reached 75 meters (246 feet), they must once more sprint half of what's left. You can keep slicing up space so that they never finish because they always have to run half of the remaining distance first. The idea of a fundamental, indivisible unit of length is one way to resolve the issue.

6.63×10^{-34}

Planck's constant $(\mathrm{J\,s})$

Early 20th-century physicists were faced with a conundrum: atoms shouldn't exist. According to what they already knew, electrons should continuously emit light as they're accelerated in their orbit around the atomic nucleus. This emission should quickly deplete the electron's energy, sending it spiraling into the center of the atom.

Packets of light

A solution began to emerge in 1900 when Max Planck suggested that light wasn't emitted continuously, but rather in discrete chunks. It is similar to shoes—you can't buy any size you like, you're restricted to predetermined choices. These subatomic chunks are referred to as "quanta," from which the field of quantum physics derives its name. Planck was not happy with the idea, however. He saw it as a last-ditch attempt to get experiment to match with theory, even referring to it as an "act of despair" in a letter to a friend.

Five years later, in 1905, Albert Einstein was able to cement the idea by successfully using it to explain a long-standing mystery known as the photoelectric effect. It is for this work—not his work on relativity—that Einstein was awarded the Nobel Prize in Physics in 1921.

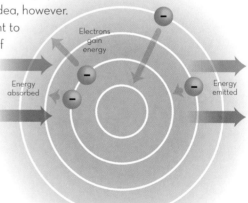

▼ By absorbing a photon, electrons can jump up an energy level. An electron falling down a level will emit a photon.

Electrons gain energy

Energy absorbed

Energy emitted

Max Planck (1858–1947)

Born in Kiel, Germany, Max Planck was a proficient piano, cello, and organ player but ultimately chose a career in physics. That was despite the advice of Munich physics professor Philipp von Jolly, who tried to warn Planck off the subject, reportedly telling him that "in this field, almost everything is already discovered, and all that remains is to fill a few holes." Famous physicist Lord Kelvin later aired similar views.

But Planck's work would explode that notion and usher in the era of quantum physics—an entirely new and often counterintuitive field. In over a century since, quantum rules have stood up to the most rigorous examination, making it one of the most precisely tested theories in the history of science. It is only by mastering this quantum world that the invention of transistors, lasers, and atomic clocks—cornerstones of our modern technological age—was possible.

Einstein said that light can only come in packets with energies that are exact multiples of the light's frequency. The gap between the energy of one packet and the next is therefore always the same. The size of that energy spacing is Planck's constant. Einstein also introduced a new word into the lexicon of physics to describe these packets of light energy: photon.

By 1913, Danish physicist Niels Bohr had applied these quantum ideas to the atom by suggesting that orbiting electrons are only permitted to have certain orbits. Electrons can jump up and down between these "energy levels" by absorbing or emitting a photon with the appropriate energy. However, once an electron reaches the lowest energy level—or ground state—it cannot spiral any further inward. The value of each energy level depends upon Planck's constant. The original conundrum was resolved.

3×10^{-34} (approx)

Wavelength of a
flying tennis ball (m)

Einstein's 1905 work on the photoelectric effect showed that light—previously considered to have purely wave-like properties—could also be thought of as made of particles called "photons."

In 1924, young French nobleman Louis de Broglie (de-BROY) wondered whether the same was true in reverse, whether objects normally seen as particles—like electrons—could also behave as waves. If a particle could behave like a wave, it would have similar wave-like properties such as frequency and wavelength.

As part of his doctoral thesis, he proposed that the wavelength of such "matter waves" could be calculated from the object's mass and velocity, as well as Planck's constant. His proposition needed to be tested experimentally to be believed, however. Any test would have to show that particles like electrons exhibit the same behavior as light waves.

Putting de Broglie to the test

If the atoms inside a material are arranged in a crystal structure, they are able to scatter a beam of X-rays (a form of light) as it encounters the crystal. If you put a screen nearby, the scattered light creates a diffraction pattern—a series of alternating bright and dark rings. Could electrons do the same?

An experiment performed in 1927 by Clinton Davisson and Lester Germer at the Bell Telephone Laboratory in New York

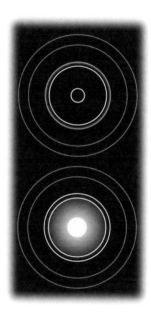

▲ The diffraction pattern formed by X-rays (top) and the diffraction pattern of electrons (bottom). This showed that matter, too, can behave like a wave.

Louis de Broglie (1892–1987)

Influential physicists normally have to wait a long time to receive a Nobel Prize as recognition for their work. Not Louis de Broglie—he picked up the accolade just five years after he first published his idea on wave-particle duality.

Born to nobility in Dieppe, France, he originally studied history with the intention of following family tradition and becoming a diplomat. However, he later switched to physics, gaining his degree in 1913. He then spent the First World War stationed in the Eiffel Tower as part of the wireless section of the French army. Returning to physics after the war, he came up with the theory for which he is most remembered.

showed that they could: the same diffraction pattern was observed—impossible if the electrons were behaving purely as particles. Matter must also be capable of behaving like a wave.

Using de Broglie's formula, the wavelength of a 57.5 gram tennis ball struck at 60 km/h comes out at 7×10^{-34} meters—an incredibly small number. The reason we don't see tennis balls (or any other macroscopic objects) exhibit wave-like behavior is that diffraction only happens if the width of the slit is similar to the wavelength of the object.

That's why sound, with its much longer wavelength, can diffract through doors, but light can't. It is also why you don't diffract through a door every time you try to enter or exit a room—a human's de Broglie wavelength at walking pace is even smaller than a fast-moving tennis ball's.

▼ Louis de Broglie showed that, like light, all matter—including tennis balls and tennis players—has a wavelength.

9.11×10^{-31}

Mass of the electron (kg)

Electrons are the workhorses of the atom. Negatively charged, they are often found orbiting around the atomic nucleus. When stripped away from the nucleus they power the modern world, streaming through wires and circuits to create electricity. The sharing or exchanging of electrons between neighboring atoms results in the diverse range of chemical compounds we see around us. By changing their positions around the nucleus, electrons can also emit and absorb light.

Yet for such important particles, they are incredibly lightweight. This is partly because, unlike protons and neutrons, they are not built of anything else. Physicists refer to electrons as fundamental.

Accurately knowing the mass of the electron is important because that value helps define other crucial numbers in physics, such as the fine structure constant (see page 50). In February 2014, a team of German physicists came up with a new method for "weighing" the electron. Their result was 13 times more accurate than any previous attempt.

Weighing anything requires a counterweight to compare against, so the German team took an atom of carbon-12 and removed all but one of its orbiting electrons. This left it with an overall negative charge, meaning it responded to the magnetic fields inside a device known as a Penning trap. The frequency with which the carbon-12 revolved around inside the trap was used to determine the mass of the overall atom—the counterweight. The mass of the electron could then be precisely inferred.

1.6726×10^{-27}

Mass of the proton (kg)

The positively charged proton is the architect of the periodic table—the elements are arranged by the number of protons they contain in their nucleus. Hydrogen has a solitary electron; rutherfordium has 104.

Unlike electrons, however, they are not fundamental particles. Instead, they are made of small building blocks called quarks. Quarks come in six different varieties called up, down, top, bottom, strange, and charm (see pages 46 and 54). A proton is made of two up quarks and one down. Their combined charges ($+^2/_3$, $+^2/_3$ and $-^1/_3$) give the proton an overall positive charge that exactly mirrors the negative charge on the electron.

However, the quark masses only account for around 1 percent of a proton's mass. The remainder has its origins in the most famous equation in the world: $E=mc^2$. Wrapped up in those numbers and letters is the idea that energy (E) and mass (m) are effectively the same thing. So any other energy associated with the constituents of the proton also contributes to its mass.

The kinetic energy (motion energy) of the quarks adds some more mass. The majority comes from the energy required to keep the quarks bound together—the two positively charged-up quarks should repel each other. However, they are "glued" together by the strong nuclear force, one of the four fundamental forces in physics, through the exchange of particles called "gluons" (see page 66). It is the energy of these gluons and their associated gluon field that dominates the overall proton mass.

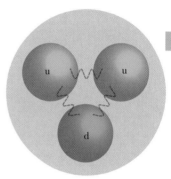

▲ The quarks in a proton are held together by the strong nuclear force. The energy of the associated gluons adds to the proton mass.

1.6749×10^{-27}

Mass of the neutron (kg)

Discovered in 1932, the neutron sits alongside protons at the heart of most atoms. While each element will always have the same number of protons, the number of neutrons it has can vary. These different versions of the same element are called isotopes. Uranium has several different isotopes, for example. The most famous are uranium-238 and uranium-235, the latter having three fewer neutrons.

Like a proton, a neutron is made of three quarks: one up and two down (as opposed to the proton's two up and one down). This gives it an overall charge of zero, hence the name. However, its mass is about 0.1 percent greater than that of its nuclear counterpart. The exact reason for this small discrepancy is currently an open research question.

Confined to the nucleus, a neutron is stable and long-lived. However, if removed from the nucleus, it will only last just shy of 15 minutes (on average). Its slightly superior mass means it can decay into a proton, an electron, and a particle called an antineutrino. Precise measurement of the average lifetime of these "free" neutrons is important in several areas of physics, including understanding how the lightest elements formed in the immediate aftermath of the Big Bang.

▲ In 1932, Sir James Chadwick (1891-1974) discovered the neutron —the neutrally charged particle at the heart of most atoms.

8.51×10^{-27}

Density of the universe (kg/m)

How do you even begin to try to work out the density of something as enormous as the visible universe? Astronomers use the overall shape of the universe as a proxy.

Throughout its history, the universe has been sculpted by a never-ending battle between gravity trying to collapse it and the outward force of its expansion. To help get a grasp on this, cosmologists define a baseline called the "critical density."

Sphere, saddle, or sheet?

If the actual density of the universe is greater than this critical density, then the universe is shaped like a sphere; less and it resembles a saddle. At, or very close to, the critical density and the universe is flat, like a sheet of paper.

By deducing which of these shapes matches the real universe, astronomers can estimate its overall density. One way to do that is to measure the angles in an enormous triangle created by the Earth and two distant objects. While angles in a triangle add up to 180 degrees on a flat surface, they add up to more on the surface of a sphere and less on the surface of a saddle.

Pristine measurements from the European Space Agency's Planck satellite suggest the universe is very close to flat. So, the density of the universe must be very close to the critical density—or an average of just over five protons per cubic meter.

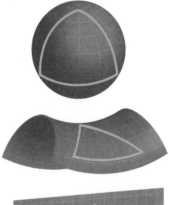

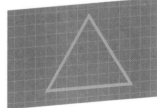

▼ Astronomers have deduced that the universe is close to flat through measuring the angles of a triangle with the Earth at one point.

3×10^{-25}

Mean lifetimes of
W and Z bosons (s)

Bosons are force carriers—particles responsible for the four fundamental forces of nature (see page 66). The weak force acts by emitting and absorbing three different bosons: W^+, W, and Z. These particles are massive, weighing more than 100 times the mass of a proton. This mass limits the force's range to just 10^{-18} meters (about 0.1 percent the diameter of a proton), so it is only felt at subnuclear scales. It also limits how long they last, explaining their tiny lifetimes.

By exchanging W bosons, quarks can shape-shift from one variety to another. The most common form of swapping is an up quark changing to a down quark (or vice versa). The Sun would not shine without this process. Our nearest star releases its energy through the process of nuclear fusion (see page 48)—a chain of reactions that ultimately converts hydrogen into helium. The first link in that chain sees hydrogen turned into deuterium, an isotope of hydrogen with one additional neutron. That neutron comes from a proton changing one of its up quarks to a down through the exchange of a W^+ boson.

The idea of a weak force mediated by W and Z bosons was first put forward by Glashow, Weinberg, and Salam in 1968. Although they shared the 1979 Nobel Prize in Physics, the W and Z bosons weren't conclusively observed until a team using the Super Proton Synchrotron experiment at CERN, near Geneva, Switzerland, found them in 1983.

▼ The Super Proton Synchrotron accelerator at CERN, the European particle physics laboratory near Geneva, was the first experiment to observe W and Z bosons.

1.38×10^{-23}
Boltzmann's constant (J/K)

What is temperature? What does it actually mean to say that one thing is hotter than another? These questions were at the forefront of physics research at the end of the 19th century, and occupied the mind of Austrian physicist Ludwig Boltzmann in particular.

The constant that bears his name is often described as a bridge between the very small and the very big, between the microscopic and macroscopic worlds. It relates the temperature of a body (such as a gas) to the energy of the individual particles within that body. Boltzmann realized that a gas with a higher temperature is simply a gas whose molecules are moving around at a faster speed. His constant helps to quantify how much the average speed of the constituent molecules increases as a gas is heated.

It most famously appears in the equation for calculating entropy—a measure of molecular randomness. In relation to our gas, it is a measure of how many ways you can rearrange the gas's microscopic molecules and not change its macroscopic properties. If there are many ways the particles can be swapped around then the system is said to have high entropy.

Boltzmann himself was plagued by mental health problems, perhaps suffering from what would be recognized today as bipolar disorder. On a trip to Italy in 1906, he hanged himself. He is buried in Vienna's Central Cemetery under a gravestone carved with the letters of his famous entropy equation, including the constant now named after him.

▲ The grave of Austrian physicist Ludwig Boltzmann (1844-1906). His famous constant is engraved on the tombstone.

1.60×10⁻¹⁹

The elementary charge (C)

First determined by US physicists Robert Millikan and Harvey Fletcher in 1909, the elementary charge is the size of the electric charge on both the electron and the proton (the electron's charge is negative). It is often denoted by a lower case "e" and is measured in coulombs. All subatomic particles have charges that come in multiples or fractions of the elementary charge, including quarks, which have charges of $\pm^1/_3$e or $\pm^2/_3$e. Neutrons have a charge of zero.

Millikan and Fletcher were able to calculate the value of the elementary charge to within 1 percent of what we know it to be today, using an oil drop experiment. A small electric charge was placed on an oil drop suspended in an electric field generated between two metal plates. The strength of the electric field was varied until the oil drop was exactly supported against gravity. Knowing the mass of the oil drop, the duo could calculate the charge on the drop necessary for it not to fall down. The experiment was repeated many times and the pair noticed that the various charges were all multiples of the same number—a number they correctly suggested was the elementary charge.

Millikan was awarded the Nobel Prize in Physics in 1923, partly for this work on the elementary charge. However, as with many Nobel awards, this wasn't without controversy. Documents uncovered after Fletcher's death seem to suggest that he was pressured into giving up his credit for the work in return for Millikan rubber-stamping his PhD.

▲ Robert Millikan (1868–1953) is most famous for his oil drop experiment, from which he was able to estimate the elementary charge.

1×10^{-18}

Density of the best
laboratory vacuum (kg/m)

If you drew an imaginary box above your head, measuring a meter (3.3 feet) on each side, the air contained within that box would weigh more than 1 kilogram (2.2lb). It would also contain 24 million million million million air molecules.

If you slowly raised the box higher above your head, the number of particles inside the box would drop, as the air gets thinner the higher you go. At 100 kilometers (62 miles)—the very edge of space—the number of particles inside the box will have dropped by a factor of 2 million. This means its density—a measure of how much mass there is in a particular volume—has also dropped.

In a laboratory, scientists can extract even more of the molecules from such a box. They can reduce the contents to just a few hundred million particles. This is the best vacuum currently achievable—the closest we can get to emptying the box completely.

However, you can never truly empty any such box. The rules of quantum physics say that, even in "empty" space, pairs of particles are spontaneously popping into existence all the time. They vanish again in an instant, but it's enough to make their presence felt. If two uncharged metallic plates are placed in a vacuum, these virtual particles nudge the plates closer together. This is called the Casimir effect, after the Dutch physicist Hendrik Casimir. Some physicists have tried to explain the accelerating expansion of the universe as the result of a similar mechanism (see page 166).

▼ Even in a near-perfect vacuum, two metal plates can be pushed together by virtual particles in what's known as the Casimir effect.

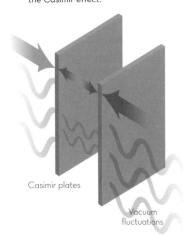

Casimir plates

Vacuum fluctuations

8×10^{-15}

Average size of an atomic nucleus (m)

Atomic nuclei vary in size greatly depending on how many protons and neutrons they contain. The smallest nucleus belongs to the hydrogen atom—it is a solitary proton just 1.75×10^{-15} meters across. Bulkier elements have much wider nuclei. Uranium-238, for example, has 146 neutrons and 92 protons, meaning it stretches to 1.5×10^{-14} meters across—approximately ten times the width of a hydrogen nucleus.

It was in 1911, while working at the University of Manchester, UK, that New Zealand-born physicist Ernest Rutherford first realized the existence of the nucleus. He was analyzing the results of an experiment that he had designed and overseen two years earlier and which was conducted by his assistants Geiger and Marsden.

Going for gold

The negatively charged electron had been discovered in 1897, but physicists knew that atoms were electrically neutral. So, they assumed there must be some undiscovered reservoir of positive charge hidden in the atom that canceled out the electron's negative charge. The discoverer of the electron, the British physicist J. J. Thomson, favored a "plum pudding model," with the electrons spread through a sphere of positive charge. Rutherford's experiment was designed to put this idea to the test.

A stream of positively charged particles was fired at a very thin sheet of gold foil. If Thomson was correct, the particles would

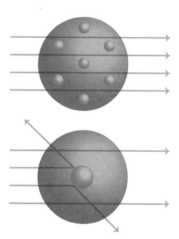

▲ If an atom's positive charge is concentrated in a central nucleus it explains the deflection of incoming particles as observed by Rutherford's experiment.

Ernest Rutherford (1871–1937)

British physicist Ernest Rutherford was already a Nobel laureate by the time his pioneering gold foil experiment was conducted by Geiger and Marsden in 1909. He had been awarded the Chemistry prize a year earlier for his work on radioactivity, although he baulked at not being given the Physics prize instead, reportedly saying that "all science is either physics or stamp collecting."

He would later "split the atom" by bombarding nitrogen atoms with alpha particles. However, he didn't quite foresee the implications of his discovery, suggesting that "The energy produced by the breaking down of the atom is a very poor kind of thing. Anyone who expects a source of power from transformation of these atoms is talking moonshine."

Rutherford died in 1937 and is buried in Westminster Abbey in London, not far from Sir Isaac Newton: apt considering that Einstein once called him "a second Newton."

have passed straight through the foil unhindered. That's not what happened. Surprisingly, some of the particles were deflected at considerable angles. Some even bounced off the foil and recoiled back toward whence they came.

Two years later, Rutherford would correctly suggest that there must be a dense packet of positive charge within atoms that was repelling any of his particles that ventured too close. The idea of the atomic nucleus was born.

That positive charge is so concentrated that the nucleus only makes up a tiny fraction of the overall atom—in a hydrogen atom it is around 100,000 times smaller than the whole. If that solitary proton was scaled up to 1 centimeter across, then the edge of the hydrogen atom would be 1 kilometer away.

8.85×10^{-12}

Permittivity of free space (C^2/Nm^2)

Permittivity is a measure of how a medium—such as air—affects an electric field. A substance with high permittivity acts to reduce the strength of any electric field present. The permittivity of free space, often denoted by ε_0, is the permittivity of a vacuum. It often crops up when looking at electromagnetic radiation, which doesn't need a medium through which to travel.

It is sometimes referred to as the "electric constant" and plays a similar role in calculating the electric force between two separated charges to Newton's Universal Gravitational Constant (see page 28) in calculating the gravitational force between two separated masses. The electric equivalent of Newton's Universal Law of Gravitation is Coulomb's Law, first formulated by 18th-century French physicist Charles-Augustin de Coulomb. The SI unit for electric charge (C) is named after him.

ε_0 also appears in two of the four famous equations known as Maxwell's equations, formulated by Scottish physicist James Clerk Maxwell in the 1860s. His work identified electricity and magnetism as two sides of the same coin, and his equations describe exactly how electric and magnetic fields are related to one another. They also explain how they originate and how electric currents and charges can affect them.

Permittivity affects how fast light travels in a particular medium. For example, water at 20°C (68°F) has a permittivity 80 times higher than free space. That leads to light traveling through that water at around 10 percent of its speed in a vacuum.

▲ Influential physicists James Clerk Maxwell (1831-1879, top) and Charles-Augustin de Coulomb (1736-1806, bottom).

$5.29{\times}10^{-11}$

Bohr radius (m)

When Ernest Rutherford discovered the atomic nucleus in 1911, he considered the atom to be an analog of the solar system: the nucleus as a stationary, central sun and the electrons as the neatly orbiting equivalent of planets.

In 1913, Danish physicist Niels Bohr altered this picture to include ideas from quantum physics. He argued that electrons cannot orbit the nucleus in any fashion they choose. Instead, most of the atom is a no-fly zone; only a handful of orbital routes—known as energy levels—are allowed. This revised picture is called the Rutherford-Bohr model, although it is more commonly referred to as just the Bohr model.

At the time it was a very powerful tool, clearing up a long-standing mystery. By the end of the 19th century, physicists had noted that atoms only emit light with certain frequencies. Bohr explained this by saying that light is emitted when an electron falls from a higher energy level to a lower one. As the gap between these levels is fixed, so is the frequency of light that such a transition can emit.

The idea of electrons in planetary-style orbits has been replaced by modern quantum physics. However, it is still a good approximation for the hydrogen atom. Bohr's model can be used to calculate the approximate distance between the hydrogen nucleus—a single proton—and an electron in the lowest energy level (the "ground state"). This Bohr radius is the rough size of a hydrogen atom.

▼ The Bohr radius is the approximate distance between the nucleus (a proton) and the single orbiting electron in the hydrogen atom.

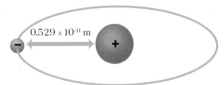

0.529×10^{-11} m

6.67×10^{-11}

Universal Gravitational Constant (Nm^2/kg^{-2})

Four centuries ago, our understanding of the universe was changing at an incredible pace. In 1609 and 1610, Galileo's observations with the newly invented telescope provided concrete evidence that the Earth orbited the Sun and not the other way around. By 1619, German astronomer Johannes Kepler had published the last of his three laws of planetary motion. His third law describes the way in which the time it takes for a planet to complete an orbit is related to its distance from the Sun. However, it was an empirical law—it was based on the observations of the planets by his colleague, the Danish astronomer Tycho Brahe. Kepler did not explain why this law should hold, merely that it seemed to fit.

▲ English physicist Sir Isaac Newton (1642-1727) is most famous for his work on gravity. His Universal Law of Gravitation was published in 1687.

Enter Newton

The breakthrough came in July 1687 with the publication of Newton's landmark work, *Principia* (see page 112). In this, in addition to his three laws of motion (see page 64), he also presented his ideas on gravity, explaining that it was the gravity of the Sun that kept the planets in orbit. He was able to derive Kepler's third law from his own Universal Law of Gravitation, lending significant weight to his ideas.

The Universal Law of Gravitation is used to calculate the gravitational force between any two masses. To find that force, you must multiply the two masses together and divide by the distance between them squared, making it a so-called "inverse square law." Double the distance between two objects and the gravitational force between them drops by a factor of four.

You must also multiply your answer by another number, known as the Universal Gravitational Constant, G. This tiny number illustrates just how weak the force of gravity is compared to the other three fundamental forces (see page 66). However, Newton was unable to calculate the value of G during his lifetime. It was eventually measured 71 years after his death by Henry Cavendish.

▲ Henry Cavendish (1731–1810) not only performed important experiments on the strength of gravity, he also discovered hydrogen.

Cavendish experiment

An accurate way to measure the value of G was first devised by British geologist John Michell, but he died before he could complete the work. His equipment eventually ended up in the hands of Henry Cavendish who, by 1798, had published his results.

The experiment consisted of a 1.8-meter (5.9-foot) wooden rod supported by a wire. A 5-centimeter (2-inch) lead ball was placed at each end of the rod. A smaller lead ball was suspended close to each of the larger balls. The tiny gravitational attraction between large and small balls caused the rod, and hence the wire, to twist. It continued twisting until the force on the wire was equal to the gravitational force between the balls. From this it was possible to calculate Newton's Universal Gravitational Constant.

What is remarkable is the level of precision required to perform such an experiment. As G is so small, so is the gravitational force between two lead balls. It is roughly equivalent to $1/50,000,000$ of the weight of the smaller balls. In order to prevent disruption by air currents, Cavendish isolated the experiment in a thick wooden box housed in a shed and used a telescope put through small holes in the side of the shed to observe the wire twisting.

1×10^{-10}

Coldest temperature
ever recorded (K)

You'd be forgiven for thinking that the coldest place in the universe would be out in space. Temperatures can plummet to 33 K (-240°C, or -400°F) on far-flung Pluto, for example. Currently, the coldest natural place known in the solar system is on the Moon. At the lunar south pole there are some craters so deep that their floors are forever shielded from the Sun. Here the temperature drops even lower than on Pluto.

Empty space itself is pretty cold, too. When astronomers look at gaps between the stars and galaxies they measure the background temperature of space to be just 2.7 K, warmed slightly by the Cosmic Microwave Background (CMB) (see page 58). There is a cosmic place even colder still. The Boomerang Nebula—a protoplanetary nebula in the constellation of Centaurus—registers at just 1 K.

▲ The Boomerang Nebula—seen here through the Hubble Space Telescope—is only one degree above absolute zero, the coldest possible temperature.

Freezing in Finland

Despite all these celestial contenders, the coldest places in the universe are created by scientists right here on Earth. The record was set in 2000 when a team from the Helsinki University of Technology, Finland, was able to cool a piece of rubidium to just 0.0000000001 K (one tenth of one billionth of a kelvin). In 2003, Massachusetts Institute of Technology set the record for a cold gas by reducing sodium to just one half-billionth of a kelvin. These super-cool materials can exhibit strange properties.

When it gets cold, things get weird

Over the past few decades, physicists have started experimenting with ultra-cool materials—those reduced to just a fraction of a degree above absolute zero. They've found that some materials behave very differently.

Bose-Einstein condensates (BEC) As a material is cooled, its constituent atoms lose energy and begin to slow down. At a critical temperature, very close to absolute zero, the atoms all pile up in the ground state. At this point the atoms effectively lose their individual identities and become indistinguishable from one another. They form a new type of matter known as a Bose-Einstein condensate, which can exhibit two peculiar properties.

Superconductivity A superconductive material is one that loses all electrical resistance when cooled to very low temperatures. Magnets made from superconducting wire are some of the most effective magnets in existence. They are used by physicists in particle accelerators, but are also found inside MRI machines in hospitals.

Superfluidity A superfluid is one that flows without friction. As all the atoms are cooled down to the point where they have lost their individual identities, they all act as one giant "super-atom." Unable to collide with each other, they don't lose energy in the same way as a normal substance. A superfluid can even climb the walls of its container.

▼ Liquid helium in its superfluid phase. Superfluidity is just one of the strange effects seen at super-cool temperatures.

Such is the interest in ultra-low temperature physics that there are plans to return the accolade of the universe's coldest place back to space. NASA plans to launch its Cold Atom Lab to the International Space Station in 2016, in the hope of achieving similar temperatures to that of the Finnish team and learning more about these unique substances.

1×10^{-8}

Amount of antimatter created at CERN (g)

Protons, electrons, and neutrons are not the whole story—the natural world has a mirrored reality at which we can only sneak the occasional glimpse. Every particle of matter has its own antimatter counterpart—a particle with the same general properties (like mass) but with an opposite charge.

Cooking up antimatter

The electron's partner is the positron, for example. This is not the same as a proton, which is much heavier. Quarks, too, have antimatter partners, allowing the existence of antiprotons and antineutrons. Physicists have been able to produce these particles artificially in machines such as the Large Hadron Collider at CERN, near Geneva, Switzerland. They've also been able to produce entire atoms of antihydrogen and antihelium. In 2011, CERN physicists were able to isolate the former for a record 17 minutes.

Antimatter does occur naturally, too. The average adult emits three positrons a minute as the result of the radioactive decay of potassium-40, which is contained in some of the food and water we consume and the air we breathe. Antimatter is also created when cosmic rays strike our atmosphere during intense thunderstorms and by certain types of medical scanners. The Sun also creates positrons as part of its chain of nuclear fusion reactions.

▲ James Cronin (b. 1931, top) and Val Fitch (1923–2015), bottom) shared the 1980 Nobel Prize in Physics for their discovery of charge conjugation parity (CP) violation.

However, compared with matter, the amount of antimatter in the universe is minuscule. All the antimatter ever created at CERN could only power a 60W light bulb for four hours. If all of CERN's high-tech gadgetry were put to work solely to produce antimatter, it would still take more than 1 billion years to amass a solitary gram of the mirror material.

The sheer effort required to produce such trace quantities makes it the most expensive material (per gram) in the world. Even the most conservative estimates put the cost of production at tens of trillions of dollars per gram. By comparison, the next most expensive material is californium-252, which comes in at an

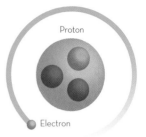

Hydrogen

Proton

Electron

Antihydrogen

Antiproton

Positron

▲ Atoms of hydrogen and antihydrogen are mirror images of one another, containing particles with the same mass but opposite charges.

Violating matter/antimatter symmetry

The simplest picture of antimatter suggests it should have been created in the same quantity as matter in the early universe. However, physicists have known since the 1960s that the two substances don't always play by such symmetrical rules.

US physicists Val Fitch and James Cronin won the 1980 Nobel Prize in Physics for their discovery of this "CP [Charge conjugation Parity] violation" during experiments at the Brookhaven National Laboratory in Long Island, New York. They were able to show that when subatomic particles called kaons decay, they unexpectedly produce unequal amounts of electrons and positrons, breaking the symmetry between matter and antimatter.

Further evidence of CP violation in different particles, known as D-mesons, was uncovered at CERN in 2011. Exactly how this difference in behavior between matter and antimatter relates to the former coming out on top in the early universe remains an active area of physics research. Physicists are also currently trying to find out whether the same discrepancy occurs in other subatomic particles, such as neutrinos (see page 63).

Antimatter as rocket fuel

The term "astronaut" translates as "star sailor," but is an enormously inaccurate description. So far, humans haven't traveled between the stars—the furthest they have reached is the Moon, about one ten-billionth of the distance to the next star after the Sun (see page 140). If we are ever to achieve interstellar travel, we need a new form of fuel.

Any propulsion system is limited by the rate at which it can propel waste products out of an exhaust. Most modern rockets carry a heavy payload of chemicals such as liquid oxygen. This is inefficient because a lot of the thrust is wasted on accelerating the heavy fuel supply. The ideal scenario would be to have the lightest fuel that packs the biggest punch. Enter antimatter. The energy created by just 10 grams (0.35oz) of annihilating antimatter could get you to Mars in a month—at least a sixfold increase in the current travel time. However, the expense of producing and containing the antimatter rules out the possibility of this for the foreseeable future.

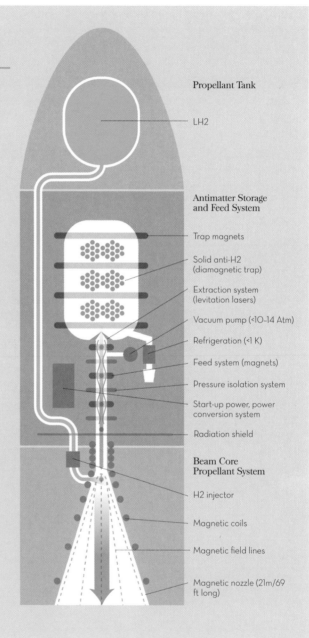

Propellant Tank

LH2

Antimatter Storage and Feed System

Trap magnets

Solid anti-H2 (diamagnetic trap)

Extraction system (levitation lasers)

Vacuum pump (<10-14 Atm)

Refrigeration (<1 K)

Feed system (magnets)

Pressure isolation system

Start-up power, power conversion system

Radiation shield

Beam Core Propellant System

H2 injector

Magnetic coils

Magnetic field lines

Magnetic nozzle (21m/69 ft long)

estimated $27 million a gram. Since the mid-1990s, gold has never reached more than $65 a gram. Despite this obvious financial hurdle, some see antimatter as the rocket fuel of the future.

Making antimatter

The main mechanism for antimatter generation is called pair production and has its origins in the equivalence of mass and energy encapsulated in Einstein's famous equation $E=mc^2$. If sufficient energy is present to cover the masses of a particle and its antiparticle—say a proton and an antiproton—then Nature can "swap" that energy for the creation of those two particles. The symmetry of this situation means that matter and antimatter should always be created in equal quantities.

But mass-energy equivalence works both ways. If, at some point down the line, a particle encounters its antimatter partner, they both turn back into energy in a process known as annihilation. The highest energy conditions in the universe's history—and therefore when most pair production took place—were right at the beginning, in the first moments after the Big Bang. Pair production eventually stopped when the expanding universe cooled below the energy required to spawn even the lightest particle/antiparticle pairs.

What's surprising is that there is any matter or antimatter left at all. In the almost 14 billion years since the Big Bang, it should have all annihilated back into energy. There should be no materials from which to build stars and galaxies. Or bones, or hearts, or DNA; life as we know it would be impossible in a purely energy-filled universe.

From the amount of matter still present in the universe today, physicists calculate that for every billion particles of matter that annihilated with antimatter, a further one particle survived the onslaught of the early universe. No primordial antimatter remains. That tiny matter surplus makes up the entire visible universe. Given their supposed symmetry, the reason for the persistence of matter over antimatter in the early universe remains one of the biggest unanswered questions in physics (see box, page 33).

5.67×10^{-8}

Stefan-Boltzmann constant

$(J\ s^{-1}\ m^{-2}\ K^{-4})$

Physicists refer to any object that absorbs all infalling radiation, regardless of the energy of that radiation, as a black body. If a black body maintains a constant temperature, the energy it emits is called black body radiation.

At the turn of the 20th century, classical physics spectacularly failed to predict the way in which the intensity of this black body radiation varied with frequency. Classical physics predicted that at higher frequencies, the intensity would go on climbing; experiments showed that, in fact, it fell. This came to be known as the "ultraviolet catastrophe."

Although not deliberately trying to find a solution, Max Planck's new idea that energy came in discrete chunks—that it is quantized—saved the day in 1900. When Planck followed the math through, his equations reproduced the observed drop-off in intensity at higher frequencies.

From this equation—known as Planck's Law—you can prove a previously known equation called the Stefan-Boltzmann Law. It shows how the energy emitted by a black body (per square meter, per second) is related to its temperature (in kelvins) and the Stefan-Boltzmann constant.

Slovenian physicist Joseph Stefan originally derived the law in 1879 and later used it to accurately calculate the surface temperature of the Sun, based on the energy it emits. It can also be used to estimate the temperature of alien planets, indicating whether liquid water is possible on their surfaces.

▲ Joseph Stefan (1835–1893) was able to calculate the surface temperature of the Sun based on the energy it emits.

4×10^{-7}

Wavelength of blue
visible light (m)

In 1905, Einstein showed that light can be treated as a stream of particles called photons. However, the idea of light as a wave dates back to the 17th century. It turns out you can think of light as both a wave and a particle—so-called wave-particle duality.

To picture light as a wave, imagine a skipping rope held at each end. If one end is wiggled up and down then a series of waves propagates down the rope. If these waves are evenly spaced, the distance between the peak of one wave and the peak of the next is known as the wavelength.

Light can take on an enormous range of wavelengths, from gamma rays with wavelengths similar to the width of an atomic nucleus to radio waves with wavelengths stretching many kilometers. However, our eyes are only sensitive to a very small window in this vast electromagnetic spectrum. We refer to that window as "visible light." The shortest wavelength of light that the human eye can pick up is around 400 nanometers (400 billionths of a meter). This appears to us as blue.

The fact that blue light has a short wavelength explains why we see the sky as that color. An object can scatter light if its size is comparable to the wavelength of the light. Four hundred nanometers is similar in size to the nitrogen molecules in our atmosphere, which means more blue light is scattered toward the ground than any other wavelength of visible light.

▼ The sky is blue because nitrogen molecules in our atmosphere preferentially scatter shorter wavelength blue light toward the ground.

7×10^{-7}

Wavelength of red visible light (m)

The longest wavelength of visible light is around 700 nanometers (7×10^{-7}); any longer than this and light strays into the infrared part of the electromagnetic spectrum.

Light with a long wavelength repeats less often and so has a lower frequency. Einstein's 1905 work showed that light with a lower frequency has lower energy (see page 114). This means light with a longer wavelength also has lower energy—red light is less energetic than blue light.

This explains the appearance of different colored flames. A blowtorch, for example, is very hot and so has high energy. The light it gives off appears blue because it has a shorter wavelength. A normal open flame is cooler, with lower energy, and so appears yellow. When a fire starts to die out and cool down it glows red because it is giving off light with an even lower energy and longer wavelength. Eventually the cooling coals no longer give out any visible light, radiating solely in the infrared instead.

The changing color of light as its waves spread out is a cornerstone principle in cosmology, too, helping Edwin Hubble discover, in 1929, that the universe is expanding (see page 132). As the universe expands, galaxies move away from us. As they do so, their light gets stretched out to longer wavelengths—they appear more red. The degree of this change in color—called redshift—is related to how fast the galaxy is receding, and therefore how fast the universe is expanding.

▲ The glowing embers of a fire appear red because as they cool they give off light of increasingly long wavelength.

1.26×10^{-6}

Permeability of free space (N/A²)

Permeability is the magnetic equivalent of electric permittivity (see page 26). It is a measure of how well a material supports a magnetic field (as opposed to an electric field). The permeability of a vacuum is often referred to as the "permeability of free space" and is known by the symbol μ_0. Highly magnetic materials, like iron, have a permeability thousands of times higher than this baseline. The term "permeability" was coined in 1885 by English engineer Oliver Heaviside. The unit of electric current—amperes—is denoted by A.

Like the permittivity of free space, μ_0 also appears in Maxwell's equations, which describe electric and magnetic fields. From his equations, James Clerk Maxwell was able to show, in 1864, that "light and magnetism are affections of the same substance." He went on to say, for the first time, that light was a form of electromagnetic disturbance.

As the permittivity and permeability of free space dictate how well a vacuum supports electric and magnetic fields, Maxwell could combine them to show how fast such an electromagnetic wave should propagate through a vacuum. His answer also matched the already accepted value for the speed of light.

Both the speed of light and the permeability of free space are currently defined by physicists to have exact values, unlike some other physical constants whose values are the result of experiment. This allows the value of the permittivity of free space to be calculated from those other two fixed values.

▲ British engineer Oliver Heaviside (1850-1925) coined the term "permeability" in the context of electromagnetism in 1885.

6.5×10^{-5}

Strength of Earth's
magnetic field (T)

The Earth has an invisible, protective envelope—a magnetic field that cocoons us from many dangers that emanate from outer space. It helps block some cancer-inducing solar UV radiation, for example. It also shields us from the worst side effects of the vicious storms that the Sun can unleash.

Ancient magnetism

And yet, despite its importance to us, the strength of the Earth's magnetic field is relatively weak. It currently ranges from 25 to 65 nanotesla (25-65 billionths of a tesla). For comparison, an ordinary fridge magnet has a strength of around 0.01T. It has been around for a long time, too. Analysis of red dacite and

▼ The Earth's magnetosphere protects us from harmful solar radiation. On the side of the Earth away from the Sun it is extended by the solar wind.

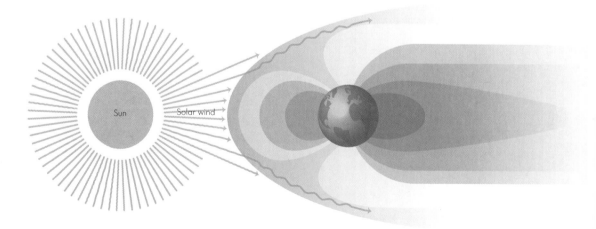

Sun

Solar wind

pillow basalt rocks in Australia suggest it has been present for at least the last 3,450 million years, around three-quarters of the Earth's current age.

It is generated deep in the Earth's core, which is separated into two parts: the solid inner and the molten outer core. Both are made of iron. The molten outer layer is highly conductive, and electric currents flowing through this region create the magnetic field.

Its sprawling magnetic fingers stretch right through the surface of the Earth and out to ten Earth radii on the side of the planet facing the Sun. Interactions with the solar wind—a stream of charged particles kicked out by the Sun—extend the field to around 200 Earth radii on the opposite side of the planet.

Incredibly dynamic, the magnetic field is also on the move—the magnetic north pole is currently migrating from Canada toward Siberia. It also varies in strength across the planet. The weakest point is known as the South Atlantic Anomaly, located over most of Brazil. The increased exposure to dangerous space radiation means the International Space Station is equipped with extra shielding to protect astronauts passing above this region.

Experiments have shown that our magnetic field is currently weakening, perhaps suggesting that we are heading for a polarity reversal where the north and south pole "flip" over. There is evidence, including from studies of stripes on the sea floor, that this has happened several times before in the Earth's history.

Flipping magnetic fields

Earth's magnetic field is weakening. Its strength has changed more in the last three centuries than it did in the previous five millennia. That could mean our protective envelope is heading for a "flip," where north and south change over, just as it did 780,000 years ago. It is worth noting, however, that bigger drops have been recorded in ancient rocks without a subsequent reversal.

So what would happen if it did flip? It is likely that its strength would only be around 20 percent of its current value. That would lead to an increase in exposure to UV radiation and the skin cancer rate would likely go up. We would also be more susceptible to damage to electrical infrastructure and satellites caused by solar storms. However, there is no evidence of any mass extinctions associated with previous flips.

1.7×10^{-3}

Rate at which Earth's rotation is slowing (s per century)

The day may be 24 hours long now, but that hasn't always been the case. Nor will it be the case in the far future. The Earth's rotation is slowing under the gravitational influence of the Moon. Billions of years ago, the day was just 18 hours long.

The Moon and the Earth tug on each other with their respective gravity. However, because the strength of gravity depends on distance, the Moon pulls strongest on the side of the Earth closest to it. This causes a "tidal bulge" on the nearside of the Earth in a direct line between the two.

As the Earth spins, it tries to carry the tidal bulge along with it. However, the Moon's gravitational pull tries to wrench the bulge

▼ The Moon's gravity is slowing down the Earth's rotation. In order to conserve angular momentum, the Moon moves away from us.

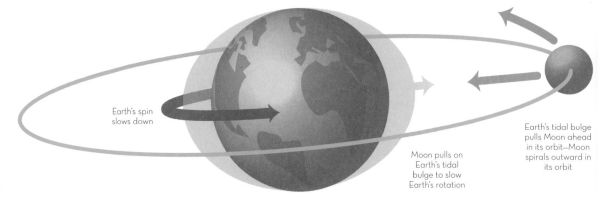

Earth's spin slows down

Moon pulls on Earth's tidal bulge to slow Earth's rotation

Earth's tidal bulge pulls Moon ahead in its orbit—Moon spirals outward in its orbit

back into place, in a direct line between it and the Earth. This tugging acts as a brake and slows down the Earth's rotation.

Moving Moon

The Earth's slowing isn't without consequence for the Moon, however. In physics, a property called angular momentum has to be conserved within a system. The effect of this is that, as the Earth loses angular momentum by slowing, the Moon acquires additional angular momentum. This boosts the Moon's orbit, moving it away from the Earth by around 3.8 centimeters (1½in) per year. This recession can be accurately measured by bouncing laser beams off a series of mirrors left on the lunar surface by visiting astronauts. Every year, the laser beams take a little longer to come back.

This slowing and receding effect—known as tidal acceleration—will continue until the rotation period of the Earth matches the time it takes for the Moon to complete an orbit. This is set to occur when the Moon is around 1.3 times further from Earth than it is now and the Earth spins once during the equivalent of 47 current Earth days. However, the death of the Sun in around 5 billion years is likely to prevent things reaching this stage.

As the Earth's rotation period is variable, it is no longer a suitable timekeeper. In fact, additional seconds—called leap seconds—are occasionally added and subtracted from our clocks so that they match the Earth's rotation. The second was once defined as a fraction of the Earth's rotation period, but the definition was changed in 1967 and is now based on radiation produced by cesium atoms (see page 131).

Leap seconds

Modern atomic clocks are now better at keeping time than the Earth. Over about 60,000 years our planet will lose a second of time, whereas the most accurate atomic clocks take 300 million years before they are out by the same amount.

However, if we relied solely on atomic time, our more accurate clocks would eventually fall out of sync with the less accurate rotation of the Earth. Far in the future, that would lead to sunrises and sunsets occurring at some pretty strange times according to our watches. So 25 leap seconds have been added in since 1972 to keep the two time systems aligned.

-273.15

Absolute zero (°C)

If you were to heat up some water, it would change from a liquid to a gas at 100°C (212°F). Continue to apply heat and the temperature of the gas will increase still further. There is theoretically no limit to how high the temperature would rise if you applied enough energy.

If you were to cool down some water, it would change from a liquid to a solid at 0°C (32°F). Continue to cool this solid and its temperature will continue to fall. However, there is a lower limit to the temperature it could reach: -273.15°C (-459.67°F). This lower limit applies not just to water but to everything in the universe. Nothing can get any colder than this.

Lowest possible temperature

Why can't temperatures ever fall below this value? Temperature is a measure of the average amount of energy that the particles in an object possess. In terms of classical physics, the temperature of an object depends on how fast its atoms and molecules are moving around or vibrating. In gases and liquids, molecules are free to move around in all directions: the higher the temperature, the faster their average speed. In a solid, atoms are held in fixed positions within a lattice structure, but are able to vibrate. The higher the temperature, the more they vibrate.

The colder an object gets, the less its constituent particles move or vibrate. Eventually, at a certain theoretical temperature,

▲ Lord Kelvin (1824-1907) and Anders Celsius (1701-1744) both have units of temperature named after them.

Celsius vs Kelvin

The Celsius temperature scale was devised such that the freezing point of water is assigned a value of zero degrees (32°F) and the boiling point of water is assigned a value of 100 degrees (212°F). This makes the Celsius scale very convenient for dealing with everyday temperatures. However, in physics, it is preferable to have an absolute temperature scale—one in which the zero value is anchored to the lowest possible temperature. Physicists use the Kelvin scale of temperature rather than the Celsius or Fahrenheit scale for this reason. The lowest possible temperature is given a value of zero degrees, after which the scale increases by the same increments as the Celsius scale. This means that 0 K is the same as −273.15°C (-459.67°F). On the Kelvin scale, the freezing point of water is 273.15 K and the boiling point of water is 373.15 K.

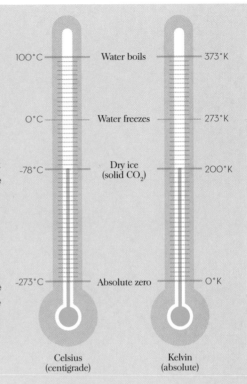

Celsius (centigrade)		Kelvin (absolute)
100°C	Water boils	373°K
0°C	Water freezes	273°K
-78°C	Dry ice (solid CO$_2$)	200°K
-273°C	Absolute zero	0°K

all movement of atoms and molecules ceases. No more heat can be removed and further cooling is impossible. This temperature is -273.15°C (-459.67°F). On the Kelvin temperature scale, this lowest possible temperature is assigned the value 0 K, and is known as absolute zero. Although classical physics predicts that all movement of atoms will cease at absolute zero, quantum mechanics predicts that, even at this temperature, atoms and molecules retain some energy, known as zero-point energy. This results in some motion even at absolute zero. In fact, in 2013, physicists in Munich, Germany, were able to cool a quantum gas below absolute zero. Such ultra-cool atoms could pave the way for the development of new materials.

$-1/3$

Charge on down, strange, and bottom quarks

(e—elementary charge)

We've already seen that protons and neutrons are not elementary particles, but are made up of three smaller particles called quarks. Quarks get their name from James Joyce's 1939 novel *Finnegans Wake*, which contains the line "Three quarks for Muster Mark." In the case of the proton, it is made of two up quarks and one down quark. The neutron is made of two downs and one up. The charges of the individual quarks sum to give an overall charge of +1 for the proton and zero for the neutron.

There are four other types of quark, however. Two of them—the strange and bottom (sometimes called beauty)—share a charge of $-1/3$ with the down quark. The other two are charm and top, which each share a charge of $+2/3$ with the up quark (see page 54).

Any particle made up of quarks is called a hadron—which is where the Large Hadron Collider near Geneva, Switzerland, gets its name. Particles that are comprised of three quarks—like protons and neutrons—are called baryons. Particles not comprised of quarks at all—like electrons—are called leptons.

There is a third alternative, however. Mesons are formed of one quark and one antiquark (an antimatter quark—see page 32). Examples of mesons that contain down, strange, and bottom quarks (or antiquarks) include the strange B meson (strange, antibottom) and the neutral kaon (down, antistrange). Mesons are unstable and so decay in less than 1 millionth of a second.

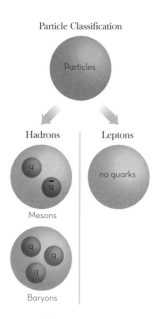

Particle Classification

▲ Particles are classified by quark content: hadrons are composed of quarks (baryons have three quarks, while a meson is made up of a quark and an antiquark); leptons have no quarks.

0

Rest mass of the photon (kg)

The photon—a particle of light and the carrier boson of the electromagnetic force—does not have any mass. At first glance, this may seem to contradict Einstein's famous equation, $E=mc^2$, which says that mass and energy are effectively the same thing. If a photon has no mass, the equation seems to suggest that it has no energy either. But light clearly has energy.

Arguably the most famous equation in the world, it isn't actually the full equation. The whole thing is $E^2=p^2c^2 + m^2c^4$, where p is the particle's momentum (which is related to its velocity). So, for stationary particles, p is zero and that half of the equation vanishes, leaving us with $E^2=m^2c^4$. Taking the square root of each side gives the equation's well-known form. For a particle with zero mass, however, the m^2c^4 bit is the half that disappears, leaving just the momentum term. That means that massless particles can still have energy as long as they have momentum.

According to Einstein's rules of special relativity, the speed of light is the maximum speed that any particle can reach. Only massless particles like photons are able to attain this maximum speed. Particles with mass can get very close to the speed of light, but would require an infinite amount of energy in order to exactly reach it (see page 100).

The zero mass of the photon can be experimentally tested. If it had even a tiny mass, then the behavior of electric fields, as summed up in Coulomb's Law (see page 26), would be affected. No such unexpected behavior has been observed.

0.007

Efficiency of hydrogen fusion

The Sun is vital to our existence. It bathes us with sufficient energy for Earth to maintain life-giving liquid water. Its energy begins every food chain across the planet. Without it, we simply wouldn't be here. Yet our local star is incredibly inefficient at generating energy. The energy per second it creates (per cubic centimeter) is less than that emitted by our own bodies. Still, it is this energy that maintains the ecosystems of an entire planet.

Deep in its core, the Sun is powered by nuclear fusion—the merging of lighter particles to produce heavier ones. There are many different forms of fusion reaction, but our star is run on the proton-proton chain. As the name suggests, the process starts with two positively charged particles. Two protons, with their matching charges, would normally repel each other. However,

▼ The proton-proton chain (pp chain) converts hydrogen into helium in the core of the Sun. Along the way mass is converted into energy.

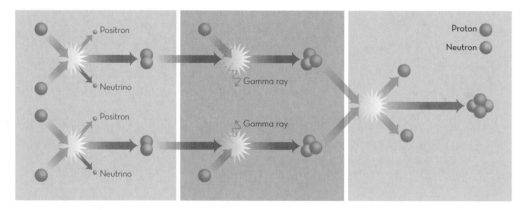

the pressure and temperature found deep in the heart of our star is immense—the core is more than ten times denser than lead. That's enough to get protons to such close quarters that they can be snared by the strong nuclear force. One of the protons then shape-shifts into a neutron to form deuterium (heavy hydrogen). Another proton joins the deuterium to form a helium-3 nucleus. When two helium-3 nuclei combine, they create helium-4—the end product of the proton-proton chain.

Four million tons of sunshine

However, there is a mismatch between the start and end of the process. Every second, the total mass of helium-4 that comes out is 4 million metric tons (around 4.4 million US tons) shy of the mass of the protons that kicked things off. Einstein famously said that energy and mass are interchangeable in his equation $E=mc^2$. That's what happens: the 4 million tons of "missing" mass is converted into the energy that helps illuminate our planet.

That may sound like a lot, but it is a small amount compared with the overall mass involved. Every second, around 620 million metric tons (around 683 million US tons) of hydrogen (equivalent to 3.7×10^{38} protons) is converted into 616 million tons (around 679 million US tons) of helium. Just 4 million tons' worth of energy is produced. That makes the efficiency of the process very low—just 0.007 or 0.7 percent. It is only the colossal size of the Sun that means the overall energy output is vast.

From the rate at which the Sun is chewing through its nuclear fuel, astrophysicists can estimate how long it will take to get through it all. It probably has 5 billion years left. When it stops fusing hydrogen, the core will begin to collapse as it is no longer supported against the pull of gravity. This gravitational contraction will raise the temperature in the core until it is possible for helium fusion to take place. In this "triple alpha process," three helium-4 nuclei end up as carbon and oxygen. This creates more energy than the proton-proton chain, and its increased outward push will cause the Sun to bloat into a red giant, perhaps even swallowing the living planet it once supported.

$^1/_{137}$ (0.0073)

The fine structure constant

In modern physics, the fine structure constant, often given the symbol α (alpha), is seen as a way to describe how strongly photons interact with charged particles like protons and electrons. As such, it has no units.

It first appeared back in 1916 when it was discussed by German physicist Arnold Sommerfeld, who first thought of it as the ratio of the speed of an electron in the first orbit of a Bohr atom (see page 13) to the speed of light.

In Bohr's model of the atom, a photon of light is emitted when an electron drops down from a higher orbit to a lower one. These emissions can be observed as a series of "spectral lines," but when you zoom in on some of the lines, it becomes apparent that they aren't single lines at all; instead, they are two lines extremely close together (referred to as "doublets"). This is the result of magnetic interactions within the atom. How close together the lines are is related to the number of protons in the atom and the fine structure constant squared.

Some astronomical studies have suggested that α isn't strictly a constant. Observations of quasars—distant, bright galaxies that appear like stars—suggest it had a different value in the universe's younger days, meaning electrons and photons could have interacted differently in the past.

▲ German physicist Arnold Sommerfeld (1868–1951) first discussed the fine structure constant in relation to the Bohr atom in 1916.

0.01

The triple point of water (°C)

In our everyday experience, water boils at 100°C (212°F) and ice melts at 0°C (32°F). But it isn't always that straightforward. At the top of Mount Everest, for example, water actually boils at 71°C (160°F). That's because the transition between the three states of water—called phase changes—also depends on pressure. At the peak of the world's tallest mountain, the atmospheric pressure is about one third what it is at ground level, so water can boil more easily.

Physicists chart this relationship between pressure and temperature on phase diagrams. The most basic of these diagrams is split up into regions showing the range of temperatures and pressures that give rise to solid, liquid, and gaseous (vapor) states. Each substance has its own phase diagram. Every diagram has a point at which the lines representing solids, liquids, and vapors intersect. At this unique combination of temperature and pressure, the substance can simultaneously exist in all three states—solid, liquid, and vapor—at the same time. This is known as the "triple point."

For water to exhibit this behavior, the temperature needs to be 0.01°C (32.02°F) and the pressure just 0.6 percent of standard atmosphere pressure. Strictly speaking, water has more than one triple point, but this is the most lauded. It is also used to define the Kelvin temperature scale. The triple point of water is defined to be 273.16 K and everything is measured from that reference point. So, absolute zero (see page 44) becomes -273.15°C (-459.67°F).

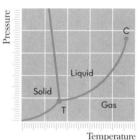

▲ At a particular combination of temperature and pressure—the triple point—a substance can simultaneously be a solid, a liquid and a gas (vapor).

0.02

The time dilation experienced
by Sergei Krikalev (s)

There can't be many of us who haven't daydreamed about the possibility of time travel. The concept has long been a cornerstone of science fiction, from H. G. Wells's *The Time Machine*, through to *Back to the Future*, *Terminator*, and *Looper*. And yet it is not some whimsical idea—it has already happened.

Einstein's work on special relativity showed us that time is fluid. Our experience of time depends on how fast we are traveling—the faster you go, the more time slows down for you (relative to a stationary observer). It is worth noting that you don't somehow feel your life passing by more quickly, as if someone hit the fast-forward button. It is just that time is passing more slowly for you *relative* to someone else's time. The bigger the difference in your respective speeds, the more you'll disagree on how much time has passed if you meet up again later.

▲ Cosmonaut Sergei Krikalev (b. 1958) has time traveled more than any other human being—0.02 seconds into his future thanks to time dilation.

Time travelers walk among us

This effect makes cosmonaut Sergei Krikalev humanity's greatest time traveler. He holds the record for the greatest amount of time spent orbiting the Earth, first on Mir and then on the International Space Station. Such craft orbit the planet at 28,000 kph (17,400 mph) relative to the ground, meaning time passes a little more slowly up there compared with down here. So, over his 803 days in orbit, Krikalev aged 0.02 seconds less than he would have had

Impossible particles: muons

When high-energy protons from cosmic rays strike our atmosphere, they create cascades of other subatomic particles, including muons. Some of these muons then travel toward the ground at 98 percent the speed of light. At this rate, they would reach the Earth's surface in around 70 millionths of a second. However, muons have a half-life—the time over which half the muons should decay—of only 1.5 millionths of a second. On paper, almost all the muons should have decayed before reaching the ground—fewer than one in a million should make it to the surface, yet experiments show that more like 50,000 survive the journey.

The big discrepancy can be explained by special relativity. At 98 percent the speed of light, the muons experience significant time dilation. Time actually runs five times slower for the muons (relative to us on the ground). Feeling as if more time is passing on their journey, fewer muons decay on their way down. Our original mistake was assuming time passes the same in their frame of reference as it does in ours.

he remained on Earth. He time traveled 0.02 seconds into his own future.

It sounds absurd, but it's true. Without accounting for time dilation, our array of GPS satellites would be useless (see page 90). We also couldn't explain the behavior of muons high in our atmosphere. One day, if we can travel at significant fractions of the speed of light, the idea of traveling years, or even centuries, into the future will no longer be the preserve of science fiction.

2/3

Charge on up, charm, and top quarks (e—elementary charge)

Along with down, strange, and bottom quarks, these three particles make up the six flavors of quark known to modern particle physics. Quarks are the only subatomic particles to experience all four fundamental forces (see page 66).

The top quark is the heaviest of all six, with a mass roughly equal to an entire atom of tungsten—almost 200 times heavier than a single proton. This enormous mass means that producing a top quark artificially requires a huge amount of energy. That's why it wasn't discovered until 1995 with the advent of modern particle accelerators. That discovery saw the 2008 Nobel Prize in Physics awarded to Makoto Kobayashi and Toshihide Maskawa, the two Japanese physicists who had correctly predicted the existence of the top and bottom quarks in 1973.

Despite what we know about quarks, no one has ever isolated one. When the top quark was discovered, it was part of a top/antitop pair (a meson). Our inability to isolate quarks is not a flaw in our imagination—it is a rule of physics. The force between quarks is referred to as the "color" force and it has some pretty strange properties. For example, as two quarks move away from each other, the force between them does not diminish. In fact, it often increases. This behavior is the complete opposite of other forces such as gravity and the electromagnetic force, whose strength drops if masses and charges are separated. The study of the forces between quarks is called "quantum chromodynamics" (QCD).

▲ Makoto Kobayashi (b. 1944, top) and Toshihide Maskawa (b. 1940, bottom), who predicted the top and bottom quarks in 1973, were awarded the 2008 Nobel Prize in Physics.

1

Thickness of graphene (atoms)

Since it was first discovered in 2004, graphene has been heralded as a miracle material that could revolutionize fields as diverse as electronics, energy storage, and medicine. At 100 times stronger than steel, it is the strongest material known to man, but at the same time is incredibly light. It conducts heat and light better than most materials in existence. All that and it is just one atom thick.

Graphene—a sheet of carbon reminiscent of chicken wire—was discovered at the University of Manchester, UK, by Andre Geim and Konstantin Novoselov. The Russian-born pair shared the 2010 Nobel Prize in Physics for their work, an unusually quick turnaround between discovery and Nobel recognition. In an age of incredibly high-tech equipment, they isolated graphene by stripping layers from a lump of graphite using adhesive tape. They were then able to successfully use their discovery in a transistor—a device used to quickly switch electric currents on and off inside computers and other gadgetry. Modern transistors use silicon, but graphene could one day replace it.

Such is the excitement over graphene's potential that, in 2013, the European Union awarded a grant of €1 billion to fund research into the area. Possible applications include tougher, more flexible touchscreens, radioactive waste management, water purification, and drug manufacture. Estimates suggest the global graphene industry could rise to $100 billion as research into the "wonder" material continues to accelerate.

▼ The chicken wire structure of graphene—a one atom thick layer of carbon—makes it one of the strongest materials known.

1.4

Chandrasekhar
limit (solar masses)

Nothing lasts forever. In around 5 billion years, the Sun will have used up its supply of hydrogen fuel. For a time, it will be able to fuse helium into carbon and oxygen, then fusion will stop forever—only stars more than eight times the mass of the Sun are massive enough to create the high core temperatures needed to fuse carbon into any heavier elements.

No longer shored against the relentless pull of gravity, the Sun's core will begin to collapse. It cannot go on collapsing forever, though. The shrinking will stop when the core reaches a size similar to the Earth. This small stellar object—called a white dwarf—will contain around half the mass of the original star (the rest having been puffed out into space, forming a planetary nebula). That amount of material crammed into a planet-sized object will make it incredibly dense—over 50,000 times denser than solid gold.

▲ The Pauli Exclusion Principle, which says that no two identical fermions can occupy the same quantum state, was formulated by Wolfgang Pauli (1900–1958).

Pauli Exclusion Principle

What stops a white dwarf collapsing any further is not the outward pressure of fusion energy, because that process will have ceased. Rather, it is something called "degeneracy pressure." In 1925, Austrian physicist Wolfgang Pauli realized that electrons play by a different set of rules from some of their fellow subatomic particles. He said that no two fermions—a class of particles to which electrons belong—can be in exactly the same quantum state. This is known as the Pauli Exclusion Principle. If the white dwarf were to collapse any further, electrons would be forced to share the same quantum state. According to the Pauli Exclusion Principle this is impossible and, therefore, the collapse must stop at that point.

In 1930, 19-year-old Indian astrophysicist Subrahmanyan Chandrasekhar realized that this means there must be a limit to the mass a white dwarf can have. White dwarfs with masses exceeding 1.4 times that of our Sun would not be stable. This restriction on a white dwarf's mass is known as the "Chandrasekhar limit." At the time, Chandrasekhar's work was largely brushed under the proverbial carpet because it implied the existence of black holes, objects a lot of astronomers were not prepared to believe in. He did eventually win the Nobel Prize in Physics, but not until 1983.

The limit that bears his name plays a fundamental role in modern cosmology. More often than not, stars exist in pairs. If one of the duo dies and becomes a dense white dwarf, it can begin to rip gas from the other. If it consumes sufficient gas to approach the Chandrasekhar limit, it becomes unstable and explodes as a type-1a supernova. As these cosmic explosions always detonate with a similar amount of fuel, they should all have a similar brightness. The dimmer the supernova appears to us, the further away it must be. That leads to type-1a supernovae being used as "standard candles," allowing us to accurately measure the distance to far-off galaxies. This exact technique was used in 1998 by two teams of astronomers to uncover the fact that our universe's expansion is accelerating (see page 96).

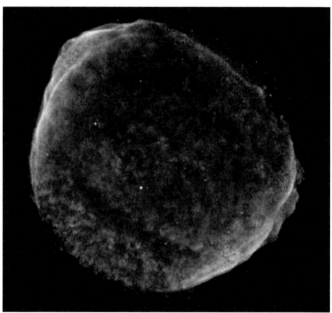

▲ SN 1006 is the likely remnant of a type-1a supernova. It exploded as a white dwarf approached the Chandrasekhar limit.

2.7

The temperature of the cosmic microwave background (K)

The theory that our universe began with a "big bang" is one of the most famous ideas in all of science. It has seeped into popular culture, too, with one of the most renowned television shows in the world named after it. Yet, despite its fame, it was only confirmed beyond reasonable doubt in the last 50 years.

The phrase itself was originally coined by one of the theory's biggest detractors, gruff British astronomer Fred Hoyle. During a BBC radio interview in 1949, he used the term to describe the main difference between it and his own favored "steady state" model. Those who trusted in the steady state theory held that the universe had been around forever; Big Bang believers subscribed to a single, violent origin. As with everything in science, differences of opinion tend to be settled by experiment.

Clues to a calamitous beginning

The idea of a "big bang" began to sprout up at the beginning of the 20th century, with larger telescopes allowing astronomers to probe deeper into the skies. Other galaxies—fuzzy patches of light then known as "island universes"—were shown to lie beyond our own Milky Way. What's more, they were moving away from us. By 1931, Belgian priest and physicist Georges Lemaître had suggested that this expansion meant that long ago all the galaxies were huddled very close together. In fact, he proposed there was a time when all the matter in the universe was concentrated into a "single quantum," which acted as the seed for our universe—the idea that Hoyle would later call the Big Bang.

Such a dense concentration of matter would have made the infant universe an incredibly hot place; far too hot for whole atoms to exist. Initially, it was too hot for even protons and neutrons to exist. Instead, for the first millionth of a second, the universe was awash with a sea of quarks and leptons (like electrons). Only after that time would the nascent universe have expanded and cooled enough for the strong force to be able to bind up and down quarks together into protons and neutrons. It was still far too hot, however, for protons to capture electrons—they simply had too much energy to be confined by the electromagnetic force. According to cosmological models, it would take around 380,000 years of expansion and cooling for electrons to finally be snared. And in this notion lies the dagger that finally killed the steady state model.

Before electrons combined with protons to make neutral atoms, the bevy of particles floating about would have kept any photons of light trapped. The photons would only have been able to move a minuscule distance before bumping into something. With electrons now crowded close to protons, photons would suddenly have had a lot more room and they would have instantaneously been able to stream outward, unhindered, at the speed of light. The universe itself continued to expand faster than the speed of light (this may seem to break the rule that nothing can travel faster than the speed of light, but that is only true of things traveling *through* space, not of the speed with which space itself can move, which has no such limitation). These ancient photons have been traveling across the expanding universe ever since.

▲ Sir Fred Hoyle (1915–2001, top), Arno Penzias (b. 1933, bottom right), and Robert Wilson (b. 1936, bottom left).

Echo of the Big Bang

Crucially, if the Big Bang picture is true, we should still be able to see these photons today. The expansion of the universe in the intervening eons should have robbed them of a lot of their energy, however, placing them in the microwave part of the electromagnetic spectrum. This reason, and the fact they should be seen in every part of the sky, led to their residual radiation being called the Cosmic Microwave Background (CMB). Its existence was predicted by Robert Dicke and George Gamow in 1946, and finding it would be a smoking gun for the Big Bang and a death knell for the steady state.

It was eventually found, by accident, in the 1960s by American duo Arno Penzias and Robert Wilson.

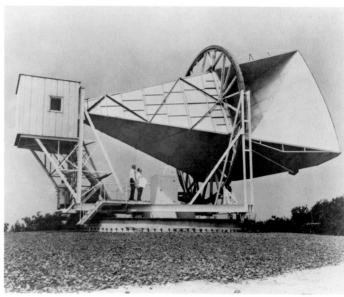

▲ The horn antenna in New Jersey, used by Penzias and Wilson to inadvertently discover the Cosmic Microwave Background (CMB).

They were experimenting in New Jersey with a radio telescope that had been designed to capture radio waves reflected from communication balloons high in the atmosphere. They ran into trouble when trying to calibrate the system by removing all other radio interference—one source persisted. They were able to rule out many origins of the nuisance signal, including human technology, the Earth itself, the Sun and anything in the Milky Way. For a time they wondered if it might be coming from a "white dielectric material" being deposited on the inside of the telescope by roosting pigeons. The pigeons were evicted and their droppings painstakingly cleared. The pigeons later returned and were shot. Despite these extreme efforts, the signal remained.

Penzias and Wilson did not realize what they had discovered. Their nuisance signal was actually the CMB. They were looking for the echo from balloons and instead found the echo of the Big Bang. Dicke eventually got wind of the discovery and realized its true significance. Despite this key role, it was Penzias and Wilson who won the Nobel Prize in Physics in 1978.

Mapping the CMB

Modern space-based instruments like the Wilkinson Microwave Anisotropy Probe (WMAP) and the Planck telescope have allowed astronomers to make exquisite measurements of the CMB. They have measured its temperature to be 2.7 K (-454.81°F). As this radiation blankets all of the sky, this is the background temperature of space. Even the voids between galaxies glow faintly at this temperature.

Locked up in the CMB are tiny temperature variations that depart from 2.7 K by about 1 part in 100,000. These indicate that the early universe had tiny density variations that gave rise to slightly hotter and cooler spots. As the universe swelled, matter gathered around the slightly denser regions. This explains the structure of the universe we see today with galaxies clumped together and separated by expansive voids. So the CMB is the key to understanding not only the universe's past, but also its present form.

▼ Modern map of the Cosmic Microwave Background (CMB) compiled using nine years of observations from NASA's WMAP satellite.

2.71...

Euler's number

Named after Swiss mathematician Leonhard Euler (pronounced "OY-ler"), this number is often represented by a lower case "e." However, the e does not stand for Euler, but for exponential. It is the rate of growth (or decay) of any continuously growing (or shrinking) process. It is an irrational number, which means its digits go on forever and it cannot be represented neatly as a fraction.

In physics, e is often used when looking at radioactive decay—it appears in the equation that tells you how many atoms of a radioactive substance you will have left after any given period of time. The average amount of time that a particle remains undecayed is also equal to the amount of time it takes for there to be only 1/e (0.368...) of the origin sample left.

It also appears when modeling planetary atmospheres, including Earth's. Atmospheric pressure decreases exponentially the further you climb from the ground. It drops by about 12 percent for every 1,000 meters of additional altitude. A quantity known as the "scale height" is the distance over which the pressure drops by 1/e. For Earth's atmosphere, this is 8,500 meters (at a temperature of 290 K, or 62°F).

Finally, it can be used to calculate compound interest. If you invest $100 at an interest rate of 5 percent, it would be worth $100 x $e^{0.05}$ or $105.13 (13c more than simple interest alone) at the end of the first year.

▲ Swiss mathematician Leonhard Euler (1707-1783) gives his name to a number important in areas concerning exponential growth or decay.

3

Flavors of neutrino

First proposed in 1930, and discovered in 1942, neutrinos are tiny subatomic particles. They get their name from the Italian for "little neutral one" and are extremely abundant in the universe. Every second there are more neutrinos passing through an area the size of your thumbnail than there are people on the planet.

However, they are also incredibly antisocial—they very rarely interact with normal matter. The fact you don't notice all those neutrinos careering through your body is testament to that. If you had a piece of lead a light year (9.46 billion million meters) long, there is a 50 percent chance of a neutrino passing straight through it.

The main source of the neutrinos ripping through you right now is the Sun. They are the by-product of the nuclear fusion in its core, produced in the stage when two protons combine to form deuterium (see page 48). Every now and again a few solar neutrinos interact with neutrino detectors on Earth. However, a mystery puzzled physicists until as recently as 2002. The number of neutrinos coming from the Sun seemed to be one third of the amount expected from theoretical calculation. It was eventually realized that neutrinos don't come in just one variety, but in three: electron, muon, and tau. What's more, they can "oscillate" between these different flavors. As the original instrumentation was only sensitive to one type, they failed to pick up on the other two-thirds.

▼ A technician tends to part of the neutrino detector housed at the Los Alamos National Laboratory in New Mexico.

3

Laws of motion in
Newtonian physics

Along with his Universal Law of Gravitation, Isaac Newton's three laws of motion are his most famous ideas. First published in his influential work *Principia* (see page 112) in 1687, they describe the relationship between forces and motion.

NEWTON'S FIRST LAW

An object that is either stationary or moving at a constant speed will continue to do so unless it experiences an external force. This is sometimes referred to as the "law of inertia"—inertia is the property of an object to resist changes in its motion.

NEWTON'S SECOND LAW

If a force is exerted on a body, the body is accelerated by an amount proportional to its mass and the size of the force applied. Force = Mass × Acceleration. This defines the "newton"—the SI unit of force. It means that 1N is the force required to change the speed of a 1 kilogram object by 1 meter per second over a period of 1 second.

NEWTON'S THIRD LAW

For every action, there is an equal and opposite reaction. This explains why objects float. The force of the Earth's gravity pulling it down (an action) is balanced by a reaction—buoyancy, an upward force exerted by the water. It also explains why we don't fall through the floor. The downwards force of gravity is balanced by the resistance of the ground: the "normal force."

▲ Along with gravity, Sir Isaac Newton (1642–1727) is most famous for his three laws of motion, published in *Principia* in 1687.

3.14...

Pi

Most famously, pi (π) is the ratio of the circumference of a circle to its diameter; the total distance around the outside of a circle that is 1 meter across will be approximately 3.14 meters. Pi is an irrational number, which means it can't be expressed as a simple fraction. The closest small fraction is $^{355}/_{113}$, which gets the first six decimal places. To ten decimal places, pi is equal to 3.1415926536. Modern supercomputers have been able to calculate pi to around 10 trillion decimal places. It has been known about for millennia.

It is not just in circles that pi crops up—it is one of the most ubiquitous numbers in all of physics and is included in a revision of Planck's constant. Normally denoted by the letter "h," Planck's constant is a fundamental part of quantum physics (see page 12), but it appears in equations alongside π so often that physicists also use the "reduced Plank constant" in their equations. Given the symbol ħ, it is equal to h divided by 2π.

Here are just some of the other numbers in this book that have π in their equation. They span quantum physics, electromagnetism, and astrophysics.

- ▶ Planck time (page 10) **and length** (page 11)
- ▶ Permeability of free space (page 39)
- ▶ Bohr radius (page 27)
- ▶ Stefan-Boltzmann constant (page 36)
- ▶ Fine structure constant (page 50)

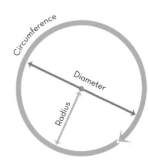

▲ Pi is the ratio of a circle's circumference to its diameter. It appears in many of physics most important equations.

4

Fundamental forces

There are four fundamental forces in nature, all operating with different strengths and dominant over different distance ranges. They are called "fundamental" because all other forces—like friction, for example—are derived from this quartet. Some only act at atomic scales, whereas others have an infinite range. Each is mediated by a particle called a boson. It is thought that all four forces were originally unified into one force at the Big Bang, before separating as the universe expanded and cooled. The sought-after framework that would unite all four forces is called the Theory of Everything (TOE).

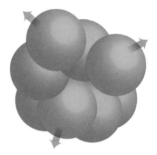

STRONG NUCLEAR FORCE

RELATIVE STRENGTH **1**	RANGE **10^{-15} m**	BOSON **GLUON**

This is the force that keeps the nucleus of an atom together. Two of the three quarks within protons and neutrons have the same electric charge, meaning they should repel each other. However, the strong nuclear force is 137 times stronger than the electromagnetic force, so it overcomes the quarks' natural desire to move apart. The same is true of the protons themselves in the nucleus. Only particles with what physicists call "color" experience the strong force. Colorless particles like electrons do not.

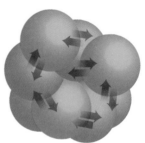

▲ Positively charged protons inside nuclei naturally want to repel each other. However, the strong nuclear force is able to trump this electromagnetic tendency.

ELECTROMAGNETIC FORCE

RELATIVE STRENGTH $1/137$ RANGE **INFINITE**   BOSON **PHOTON**

Along with gravity, electromagnetism is the fundamental force that we have most experience of in our day-to-day existence. It was James Clerk Maxwell who, in 1873, realized that electricity and magnetism were actually two aspects of one underlying force. The photon was discovered in 1905 by Albert Einstein during his work on the photoelectric effect. Like gluons, photons are massless (see page 47). At high enough energies, the electromagnetic force combines with the weak nuclear force into the electroweak force.

WEAK NUCLEAR FORCE

RELATIVE STRENGTH **10^{-6}**   RANGE **10^{-18} m**   BOSON **W+, W- AND Z**

This force derives its name from the fact that it is 1 million times weaker than its strong nuclear counterpart. However, it is still many million times stronger than gravity. Its existence was first revealed during the study of radioactive decay, in which it plays a fundamental role. In beta minus (β^-) decay, for example, the weak force acts to transform a neutron into a proton, an electron and an antineutrino. It is a "virtual" W- boson that decays into the electron and the antineutrino. As the ejected electrons move at high speed, this radiation can be very dangerous. However, in controlled environments it can be used to treat certain forms of cancer.

GRAVITY

RELATIVE STRENGTH **6×10^{-39}**   RANGE **INFINITE**   BOSON **(GRAVITON)**

Gravity is the outcast of the four fundamental forces—it doesn't seem to play by the same rules as its compatriots. It is considerably weaker than the other three forces, for example. The exact reason for this is still open to debate (see page 158). While the other three forces also have confirmed bosons, the graviton is still a hypothetical particle. At present, physicists cannot get gravity to fit into the framework of quantum physics. Many attempts, including string theory, have been made to bring gravity into the quantum world, so far without success (see page 76).

4

Dimensions in spacetime

In our everyday experience, space and time seem like two separate concepts. For a start, there are three ways that you can move in space: up and down, left and right, and back and forth. If you wanted to tell me your exact location on the Earth, you'd need to state your latitude (how far up or down from the equator you are), your longitude (how far from the Prime Meridian at Greenwich you are), and your altitude (how high above sea level you are). Your place in time requires only one such coordinate.

Time also appears to have a direction. It moves forever onward, marching us from the past to the future. You can retrace your steps in space; you can't in time. Yet today we know the two seemingly disparate concepts are both part of a single entity, bundled up together in an all-pervading four-dimensional fabric known as spacetime.

▲ Hermann Minkowski (1864-1909) coined the term "spacetime" when discussing Einstein's work on special relativity.

Time warps

Although others had contemplated this marriage before, it is Albert Einstein who is most associated with it, thanks to his theories of relativity. The term was coined by Hermann Minkowski in 1908 when discussing Einstein's 1905 work on special relativity. Einstein's later work on general relativity, published in 1915, implied that gravity is not some mysterious attraction between two bodies, but the result of massive objects curving spacetime. The conventional way to picture this warping is by imagining a sheet of plastic held tight at each corner, with a bowling ball placed in the center to represent the Sun. This causes the sheet to droop in the middle. If a smaller object, say a marble, is rolled fast enough around the rim of this "gravitational well," it will stay in orbit around the bowling ball. This is how Einstein thought of gravity, simply a manifestation of massive objects sagging spacetime. The idea solved a long-standing mystery regarding the orbit of Mercury and was later confirmed through observations of a solar eclipse in 1919 (see page 92).

The warping of spacetime by massive objects doesn't just affect space: it affects time, too. Einstein's work showed that time runs at different rates depending on how far into a gravitational well you find yourself. This means that highly accurate atomic clocks will eventually fall out of sync with others kept on shelves of different heights, even in the same laboratory—the lower clock runs slower because it sits slightly deeper in Earth's gravitational well. It also means clocks on-board orbiting satellites—which sit higher up in the well—run faster than those on the ground. If this discrepancy were left uncorrected, our GPS satellites would be useless (see page 90).

Taking the idea to its extreme, it is possible to imagine a place where time is seen to slow down so much it stops. This is exactly what an observer would see as an object approached a black hole—an object so massive that nothing can escape the clutches of its gravitational well, not even light.

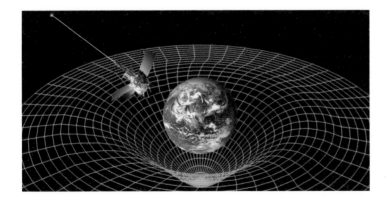

◀ Clocks aboard satellites in orbit speed up compared with clocks on Earth because they sit in a weaker gravitational field.

4.186

Joules in a calorie

Barely a day goes by when we aren't confronted by calories. Whether on television, in newspapers and magazines, or on the back of food packets, the unit of energy is a common feature of modern life. The tangled web of history means that there are actually two main definitions of a calorie, which can lead to confusion when it comes to nutrition. There is the small calorie (with a lower case "c"), which is the energy required to raise the temperature of a gram of water by 1°C (33.8°C). There is also the Calorie—the energy required to raise the temperature of a kilogram of water by the same amount. So one Calorie is equal to a thousand calories, which is why "kcal" often appears on food and drink packaging.

To make matters worse, there are also many types of calorie in science. The most widely used is the thermochemical calorie, the exact equivalent of 4.186 joules of energy. As one joule is one newton-meter, and therefore derived from SI units, it is preferable to the calorie.

The amount of heat required to increase the temperature of a substance by 1°C is called the "specific heat"; the specific heat of water is equal to 4.186 kJ/kg°C. This is higher than most common substances and explains why the oceans can hold on to a lot of the heat from the Sun.

▲ The unit of energy is named after British physicist James Joule (1818–1889). He also worked with Kelvin on temperature scales.

4.23

Distance to nearest star
(light years)

The meter is a perfect choice of distance unit when it comes to life on Earth. Like the second—which is approximately the length of a human heartbeat—it fits our everyday experience. Most humans are between 1.5 and 2 meters (between 5 and 6½ feet) tall, so it's an easy unit to picture. However, just as it doesn't pay to write the 13.8 billion year age of the universe in seconds, it is tedious to write the vast distances between the stars in meters. Or kilometers, for that matter.

Astronomers instead define a new unit: the light year. Although it sounds like a measure of time, it is the distance covered by a beam of light in one year. Traveling at 299,792,458 m/s, a light year is then 9.46×10^{15} (or 9.46 billion million meters). The nearest star to the Earth (after the Sun) is Promixa Centauri and, being 4.23 light years away, it takes the light from that star 4.23 years to reach us.

As light is the fastest thing in the universe (see page 126), nothing can cover that distance any quicker than 4.23 years. If we tried the journey with the fastest rockets ever designed, we still wouldn't even come close. In fact, it would take us around 60,000 years to make the trip. That's compared to a three-day trip to the Moon or a six-month jaunt to Mars.

▼ The nearest star to us after the Sun, Proxima Centauri would take around 60,000 years to reach at current rocket speeds.

6

Planets known to the ancients

In the centuries, even millennia, before the invention of the telescope, humans already knew of five planets in addition to the Earth.

As the Earth orbits the Sun, we see different constellations wheel through the night sky. The constellations move as a whole, staying in fixed patterns that return year after year. Civilizations including that of ancient Greece noted that the Sun, Moon, and five other "wandering stars" traverse their way through the constellations. The Greek for wandering star is "asteres planhtai," from which we get the modern word "planet." Today, we know those five wanderers as Mercury, Venus, Mars, Jupiter, and Saturn.

Saturn was the most distant planet known until March 13, 1781, when William Herschel discovered the planet Uranus through his telescope from his townhouse in Bath, England. This immediately doubled the size of the solar system, as Uranus orbits twice as far from the Sun as Saturn.

As Uranus was studied further, astronomers noted peculiarities in its orbit: it seemed to be slowing down and speeding up. It was suggested that an eighth planet might lie even further out, and its gravity was pulling Uranus around. When astronomers pointed their telescope at the predicted location of this new world in 1846, they found Neptune waiting for them. A ninth planet—Pluto—was discovered in 1930, but downgraded to a "dwarf planet" in 2006, mainly because it crosses orbits with Neptune.

▲ Sir William Herschel (1738-1822) effectively doubled the size of the solar system when he discovered the planet Uranus in 1781.

8.314

Ideal gas constant (J/mK)

In physics, it often pays to develop a simple, more manageable picture of what is going on. It saves overcomplicating things when a deeper picture is not necessary. The simplest picture of a gas—known as an ideal gas—is a good example.

A gas is made up of lots of molecules all moving around and bumping into both each other and the walls of whatever contains it. Pressure is the result of these collisions. In an ideal gas, the molecules always travel in straight lines and are pictured like snooker balls. It is assumed that these rigid spheres don't slow down when they hit a neighboring molecule or the wall of the container (such collisions are described in the jargon of physicists as "perfectly elastic"). It is also assumed that there are no other interactions between the molecules.

Such idealized behavior allows physicists to relate the pressure and volume of a gas to its temperature. In this equation, the ideal gas constant links those three properties to the number of moles in the gas, where one mole contains a total number of molecules equal to Avogadro's constant (see page 149).

While this picture works well for most everyday gases, it begins to break down if the pressure or temperature is greatly varied. One reason for this is that there are non-collisional interactions between gas molecules called van de Waals forces.

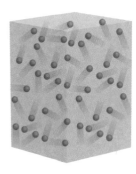

▲ The particles in an ideal gas travel in straight lines and don't slow down when colliding with each other or their container.

9.81

Acceleration due
to gravity $(\mathrm{m/s^2})$

In terms of apocryphal tales from the scientific history books, Galileo dropping cannonballs from the Leaning Tower of Pisa is right up there with Newton being struck by a falling apple. While it may not have happened, Galileo certainly did work on falling objects.

The Italian astronomer and mathematician suggested that any two objects are accelerated toward the ground by the Earth's gravity at the same rate. It doesn't matter if you drop an elephant, a cannonball or a feather—each will undergo the same acceleration. The object's mass and shape don't come into it. On Earth it doesn't quite play out that way, because of air resistance. A feather is less aerodynamic and so will float down gently; a cannonball will plummet.

But in a vacuum, Galileo's idea holds true. A perfect demonstration was performed on the surface of the Moon by Apollo 15 astronaut Dave Scott in 1971. He dropped a hammer and a feather from the same height and, with no atmosphere to resist their motion, they hit the lunar dust simultaneously.

On Earth, and at sea level, the acceleration due to gravity (g) is defined as exactly 9.80665 $\mathrm{m/s^2}$. In reality, it varies over the surface of the Earth. The value of g depends on how much mass there is underneath you and how far you are from the planet's center. The position of ocean trenches and mountains, along with variations in the planet's density, make it more complicated. Earth's complex gravity field has been mapped by satellites like

▲ Italian astronomer Galileo Galilei (1564-1642) revolutionized our understanding of the universe with his telescope, and also theorized about falling objects.

the European Space Agency's Gravity field and steady-state Ocean Circulation Explorer (GOCE) mission, which ended in November 2013.

▲ Colonel John Stapp (1901-1999) enduring forces up to 22 times the Earth's gravity (22g) during a trip on a rocket-propelled research sled.

Weighing you down

Despite the fact we use the term "weight" all the time in everyday language, it has a very specific meaning in physics. Your weight is dependent on the acceleration you feel due to gravity—how much the object you are on is pulling you down. Your mass, however, is the total amount of "stuff" you are made of and doesn't change in different gravitational fields. As it is a force, weight is measured in newtons. Mass is measured in kilograms. To calculate your weight, you multiply your mass by the appropriate acceleration due to gravity. So the weight of a 70 kilogram person at sea level would be 70 x 9.80665 = 686.5N. However, at the top of Mount Everest, the same 70 kilogram person would weigh slightly less. At nearly 9 kilometers above sea level, g at the summit is around 9.77 m/s^2, resulting in a weight of 684.6N. Their mass would still be 70 kilograms.

Another related term is "g force," which often features when discussing rollercoasters, fighter pilots, or racing car drivers. If you undergo rapid acceleration, your weight increases and you feel heavier as you get pushed into a surface (like your seat). This additional acceleration is measured in multiples of g. In a fighter jet, pilots can experience up to 12g; they require specific training to allow their body to cope with such extreme forces.

▼ The Leaning Tower of Pisa in Italy. Galileo is reported to have dropped cannonballs from it, although this probably never happened.

11

Number of dimensions
proposed in M theory

The ultimate goal for many theoretical physicists is to find one overarching theory that is capable of explaining all phenomena in nature, from the smallest subatomic particle to the structure of the universe on the biggest scales. That would require uniting all four fundamental forces (see page 66). Electromagnetism and the weak nuclear force have already been unified into the electroweak force. At high enough energies, the behavior of both can be seen as manifestations of one overall force. It is believed that when the universe was just one unit of Planck time old (see page 10), the strong force was also unified with the electroweak force. Gravity doesn't appear to play so nicely.

Attempts to unify gravity with the other three forces normally see the equations explode in a sea of infinities. However, an idea known as "M theory" at least manages to unify gravity with the other forces on paper. M theory has its origins in string theory, which says that particles like quarks and electrons are not the most fundamental building blocks of atoms. They propose instead that they are made up of tiny vibrating strings. Just like plucking violin strings in different ways results in different musical notes, so string theorists argue that Nature makes different particles by "playing" these subatomic strings in different ways.

When these ideas are applied to the task of uniting gravity with the three other fundamental forces, it works. The equations don't blow up. There is a catch, though: the math only works if our universe contains 11 dimensions (ten of space and one of time). That's a full seven more spatial dimensions than we experience on a day-to-day basis. Where are these hidden dimensions? Why don't we see them?

Hidden dimensions

One explanation was put forward long before the idea of string theory was even born. Swedish physicist Oskar Klein faced a similar dilemma in the 1920s. A German mathematician called Theodor Kaluza had attempted to unite gravity with electromagnetism. He was able to do it on paper, but needed to invent one additional dimension of space in order to get his sums to work. Klein suggested that if this "missing" dimension was curled up really small—about the same size as the Planck length—then we wouldn't notice it. This idea came to be known as Kaluza-Klein theory, but there is no evidence our universe really has this extra dimension.

However, when string theorists realized they needed 11 dimensions to get their math to work, they turned to the same explanation as Klein—that they are wrapped up so small as to remain undetected. What's unclear is whether these additional dimensions are real. At present, there is no experiment that can be performed to find out, so the situation is tantalizing. We might already have our hands on an all-powerful theory that generations of physicists have dedicated their lives to finding. Or it might all be a beautiful mathematical red herring. For now, the search for unification of Nature's four fundamental forces continues.

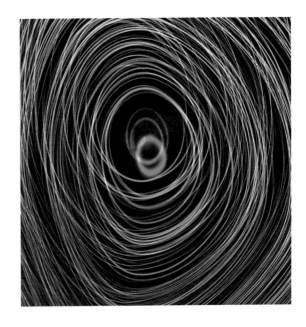

▲ If string theory is true, then particles like electrons, previously thought to be fundamental, are actually made up of tiny vibrating strings.

11.2

Earth's escape velocity (km/s)

Space isn't that far away. It begins just 100 kilometers (62 miles) above your head—a distance that many people drive on a regular basis. And yet only around 500 people have ever made it there. That's because it is incredibly hard to get to, and Earth's gravity is to blame.

In order to escape our planet's gravitational grasp, your kinetic energy must exceed the gravitational potential energy of the Earth at its surface. This requires a speed of 11 km/s (6.8 m/s). This is an instantaneous speed—the speed you need to launch at without putting in any additional energy as you go. Technically, you could get into space traveling at just 1 m/s. To maintain this speed, however, you would have to supply more and more energy to prevent gravity from slowing you down. Launch at an object's escape velocity and it cannot slow you down completely before you've escaped.

No hydrogen, no helium

This number explains why our planet has no hydrogen and helium in its atmosphere. They are the two lightest elements in the universe and can be accelerated to high speeds by energy from the Sun. Once they reach speeds greater than the escape velocity, they are lost to space forever. Thankfully, heavier elements like oxygen very rarely reach such speeds and our planet maintains its life-giving atmosphere. Earth is still

▲ The Space Shuttle *Endeavour* takes off on a mission from the Kennedy Space Center in 2009, reaching Earth's escape velocity along the way.

losing hydrogen, however. As water evaporates it rises into the atmosphere, where some of it is broken down into hydrogen and oxygen by ultraviolet radiation. Any hydrogen subsequently reaching the escape velocity is lost.

The escape velocity depends only on the mass of the object being launched from, not the object being launched. So the escape velocity of the other objects in the solar system depends only on how massive they are. The escape velocity of the Moon is 21 percent that of Earth, meaning it requires less energy to launch from. This has led to calls for a future permanent lunar base. It would be easier (and cheaper) to explore the solar system that way because you wouldn't have to keep launching from the bulky Earth. You could return to the Moon to restock and refuel before heading back out again.

Mars's escape velocity, on the other hand, is almost exactly twice that of the Moon (about 45 percent of Earth's). That is one of the major hurdles currently preventing humans setting foot on the Red Planet. Unlike the rovers we've sent to explore the planet, humans require a return journey. Taking the necessary hardware to attain Mars's escape velocity is no mean feat and currently beyond our technological capabilities.

The solar system itself also has an escape velocity. Objects like comets and asteroids can be ejected entirely if they are accelerated up to the Sun's escape velocity of 617.5 km/s (384 m/s). For even more massive objects—black holes—escaping is not possible. The escape velocity of a black hole exceeds the speed of light—the fastest possible speed. Once you're in, you're in.

▼ Mars is less massive than Earth, so it has a lower escape velocity, about 45 percent of Earth's, but this is still enough to make returning difficult.

22

Focal length of
the human eye (mm)

The eye is one of the most intricate parts of the human body. Eyes play such a fundamental role in the survival of so many species they have cropped up independently many times in the course of Earth's evolutionary history. Biologists call this convergent evolution.

Eyes like cameras

There are generally considered to be ten types of eyes in nature. Our particular type of eye is called "camera type"—the light we see ends up being focused onto the retina at the back of our eye, much like light is focused onto a camera's CCD chip. Light first

▼ The human eye uses a lens to focus an inverted image on the retina at the back of the eye.

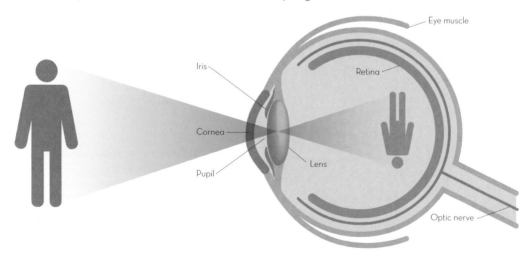

Eye muscle

Iris

Retina

Cornea

Lens

Pupil

Optic nerve

enters our eyes through the cornea—a thin, transparent layer at the front of the eye. The cornea has a refractive index of 1.38, meaning light passes through it 1.38 times more slowly than it does through a vacuum (or air for that matter, which has a refractive index of 1.000293). This results in the light being bent toward the pupil. After entering the pupil, the light encounters the lens, which is held in place by the ciliary muscles. After being focused by the lens, the light subsequently passes through a jelly-like substance called the vitreous humor, before striking the retina. Light-sensitive cells called rods and cones convert the light into electrical signals, which are then ferried to the brain via the optic nerve. The brain interprets these signals as what we see around us.

Any optical system, whether it be a telescope, camera, or human eye, has a property called the "focal length." This is the distance over which light rays are brought to a focus. With a relaxed eye, this is around 22 millimeters ($^{7}/_{8}$in). The shorter the focal length of an optical device, the more it bends the light rays. This results in focus being achieved over a shorter distance. For observing close objects, the ciliary muscles slacken and the lens becomes fatter. This increases the eye's focal length. For more distant objects, the muscles contract and the lens becomes thinner, resulting in a reduced focal length.

This ability to rapidly change between viewing near and far objects is known as "accommodation." While it is possible to do this consciously, it is almost always a reflex and a very quick one at that. A young, healthy eye can switch from focusing on a point far away to one just 7 centimeters ($2^{3}/_{4}$in) away in about one third of a second.

Short vs long sight

Shortsightedness (myopia) is often caused by an elongated eyeball, meaning that the retina is placed further away from the lens. This means that the image comes to a focus in front of the retina so all the retina can pick up is a blurred image. Longsightedness often occurs in old age when the ciliary muscles weaken and can no longer manipulate the lens into the right shape to achieve the appropriate focal length. Both conditions can be corrected by placing a lens in front of the eye in the form of glasses or contact lenses.

24

Number of elementary particles in the Standard Model

The Standard Model of particle physics describes the structure of subatomic particles and their interactions via the strong, weak and electromagnetic force. At the moment, it contains 24 fundamental particles (12 particles and their 12 antiparticles). A fundamental particle is one that is not built from anything smaller—a proton is not fundamental because it is made of quarks.

ELECTRON AND POSITRON The electron binds to the positively charged nucleus to make an atom electrically neutral. It was discovered in 1897 by J. J. Thomson. Its antimatter counterpart was discovered in 1932 by American physicist Carl Anderson.

TAU (AND ANTITAU) The tau is nearly 3,500 times heavier than the electron, but they share a charge of −1. With a mean lifetime of just $2.9 \cdot 10^{13}$ seconds, the tau are only observable in high-energy situations such as those found inside particle accelerators. It was discovered in 1975 by American physicist Martin Perl.

MUON (AND ANTIMUON) Discovered by Carl Anderson in 1936, the muon is over 200 times heavier than the electron, shares the same charge and has a mean lifetime of 2.2 microseconds.

ELECTRON NEUTRINO (AND ELECTRON ANTINEUTRINO) This is produced by nuclear fusion in the Sun's core. Two-thirds of solar neutrinos change into muons and taus en route to Earth.

MUON NEUTRINO (AND ANTIMUON NEUTRINO) Like all neutrinos, it is electrically neutral. It was discovered in 1962 by American physicists Leon Lederman, Melvin Schwartz, and Jack Steinberger.

TAU NEUTRINO Although Perl's discovery of the tau particle implied the existence of a tau neutrino (the electron and muon already had their counterparts), it wasn't actually discovered until July 2000 by a team of physicists at Fermilab in Illinois.

UP QUARK (AND ANTIUP QUARK) The up quark was first proposed in 1964 and discovered in 1968. Along with one down quark, two up quarks form positively charged protons.

DOWN QUARK (AND ANTIDOWN QUARK) The down quark was discovered by the same team and in the same year as the up quark. A neutron is made from two down quarks and one up quark.

TOP QUARK (AND ANTITOP QUARK) Sometimes called the "truth" quark, it has an electric charge of $+2/3$ like the up quark. When it decays, it almost always turns into a W boson and a bottom quark. It wasn't discovered until 1995.

BOTTOM QUARK (AND ANTIBOTTOM QUARK) With a charge of $-1/3$, the bottom quark is over four times heavier than a proton. It was discovered in 1977 by a team led by Leon Lederman.

STRANGE QUARK (AND ANTISTRANGE QUARK) Although it was officially discovered in 1968, a particle containing a strange quark (a kaon) was actually found back in 1947.

CHARM QUARK (AND ANTICHARM QUARK)
In 1974, two separate teams discovered a particle called the J/ϕ meson (a meson comprises a particle and an antiparticle). The J/ϕ meson is made of a charm/anticharm pair and it was the first time these particles had been identified.

26.8

Percentage of the universe thought to be dark matter

In March 2013, the team behind the European Space Agency's Planck telescope released the most accurate map of the Cosmic Microwave Background (CMB) (see page 61) ever produced. From its lookout some 1.5 million kilometers (930,000 miles) from our planet, the telescope had started capturing these ancient photons shortly after its launch in 2009. CMB maps from previous experiments revealed tiny speckles—regions where the temperature varies from

▼ The IceCube Detector—the world's largest neutrino observatory—is just one of the ways physicists are hunting for dark matter.

Hunting for WIMPs

While we can't see Weakly Interacting Massive Particles (WIMPs) directly, it is still possible for them to make themselves known through interactions with each other. When two WIMPs meet, they should annihilate into a cascade of other particles, some of which—like neutrinos—we should be able to pick up.

Due to its density, one of the most likely places for WIMP interaction is in the center of our Milky Way. The AMS-02 experiment, strapped to the International Space Station, is currently looking for an excess of positrons coming from the heart of the galaxy. What's been found so far is promising, but more data are still required to know for sure.

WIMP interactions could be happening closer to home, too. As the Sun sweeps through the galaxy, it should be a magnet for WIMPs. As they meet in the Sun, they should annihilate. Most of the by-products would never make it out of our densely packed star. But neutrinos, with their reluctance to interact, would make it out and telescopes like IceCube stationed on the ice plains of Antarctica should be able to pick them up.

the background by a minute slither of a degree (about one part in 100,000). Now the size and location of these speckles is known more accurately. Imprinted in these tiny fluctuations is information about the ratio of normal matter (called baryonic matter) to dark matter. Planck showed that our universe comprises 26.8 percent dark matter and just 4.9 percent atoms. The remaining 68.3 percent is thought to be dark energy (see page 167).

Hints at hidden material

Clues to the existence of this mysterious dark matter started to spring up as early as the 1930s. Swiss-American astronomer Fritz Zwicky was looking at clusters of galaxies when he found one—the

Coma Cluster—that seemed impossible. It simply shouldn't exist. He was able to measure the speed with which the galaxies in the cluster were buzzing around. According to our understanding of gravity, some of these fast-moving galaxies should have reached the escape velocity (see page 78) of the cluster and been ejected. Yet there they were, bound in a group. Zwicky was confronted with two possible explanations: either the rules of gravity weren't the same in the cluster as they are here on Earth, or there was something else lurking in the cluster that he wasn't able to see. The gravitational pull of this "missing" mass would increase the escape velocity of the cluster and so the galaxies could whizz around at high speed but still remain bound. He referred it as

Fritz Zwicky

There have been few astronomers in history more eccentric than Fritz Zwicky. Born in Bulgaria on Valentine's Day 1898, he moved to Switzerland aged six to live with his grandparents. In 1925, he moved to the United States to work with Robert Millikan—he of the famous oil drop experiment (see page 22).

Working on stellar evolution with Walter Baade, he was one of the first to grapple with the physics of neutron stars. He also the coined the word "supernova." Remarkably forward-thinking, he realized that the predictable nature of supernovae could be used to probe distances in space. This technique would eventually lead to the discovery of dark energy in 1998 (see page 166).

Not all of his ideas worked out, however. In an attempt to reduce the turbulent air around his telescope, he had his assistant fire a gun out of the slit. The turbulence remained.

A bit of a loner, he didn't always get on with his colleagues. One of his most famous put-downs was to call someone a "spherical bastard"—meaning a person who is equally unpleasant no matter which way you look at them.

"dunkle Materie"—the German for dark matter. There had to be around 400 times more mass in the cluster than he could see with his telescopes.

Soon, a similar phenomenon was observed in our own galaxy by Dutch astronomer Jan Oort. He was measuring the speed of stars orbiting near the edge of the Milky Way, and again found them moving too fast to be kept in orbit by the gravity of all the visible matter in the galaxy. However, despite these hints, the idea of dark matter wasn't taken seriously until the beginning of the 1980s. By that point, American astronomer Vera Rubin had seen the same "impossible" stars in around 100 other galaxies. Theorists then turned to what this mystifying material might be.

Certain things were clear. It had to interact with normal matter via the gravitational force, but it couldn't interact via the electromagnetic force, otherwise we'd be able to see it as it reflected photons. The trouble is that there is nothing in the Standard Model that behaves like this—none of the 24 elementary particles (see page 82) are up to the job. Currently, the most currently favored explanation is that dark matter is made up of as yet undiscovered particles called Weakly Interacting Massive Particles (WIMPs). The hunt is on to find them (see box, page 85).

However, some physicists have argued that we don't need a new particle to explain dark matter at all. They return instead to the other possibility facing Zwicky in the 1930s: that gravity isn't the same everywhere. This notion goes by the name of Modified Newtonian Dynamics (MOND). Our best theory of gravity is Einstein's general theory of relativity, and it perfectly explains the orbits of the planets and even cleared up a long-standing mystery related to Mercury's orbit (see page 68). However, there is yet no direct evidence that the same rules apply on bigger galactic scales.

27

Length of the Large
Hadron Collider (km)

Geneva is a fascinating and diverse city. Situated near
the French/Swiss border, it is home to many international
organizations. The Red Cross, The World Health Organization
and Médecins Sans Frontières are all based there. It is also where
the European Organization for Nuclear Research (CERN) chose
to set up shop in 1954.

Thanks to the recently completed search for the Higgs boson,
its major experiment—the Large Hadron Collider (LHC)—has
become a household name. Sandwiched between Geneva
airport and the Jura Mountains, its 27 kilometer (16.7 mile) long
circular tunnel is buried 100 meters (328 feet) underground. This

◀ The path of the Large
Hadron Collider tunnels
under the French-Swiss
countryside close to Geneva.
Geneva Airport can be seen
in the foreground.

subterranean scientific leviathan is the world's largest particle physics laboratory.

Over 10,000 scientists from more than 100 countries are involved in the operation of the LHC, which was constructed between 1998 and 2008. Over 1,600 giant, supercooled magnets direct two beams of protons in opposite directions around the extensive track. The protons can reach speeds incredibly close to the speed of light, before smashing together with such force that a sea of daughter particles is created for the physicists to pore over.

Beyond the Standard Model

While the Higgs boson has now been ticked off the list, there are still many outstanding mysteries that it is hoped the LHC can answer. Of particular interest is why there was a discrepancy between matter and antimatter in the early universe—a discrepancy that allows our very existence. Evidence for theories beyond the Standard Model—such as supersymmetry—is also being sought. Supersymmetry (SUSY) posits that every particle has a supersymmetric partner (or sparticle). The quark would have a particle called a squark, the neutrino a sneutrino. It is thought that the lightest of these SUSY particles could be the elusive WIMP (see page 84), held responsible for dark matter.

The LHC's four main detectors

The LHC has a total of seven detectors. Below are details of the four major ones.

ALICE Standing for A Large Ion Collider Experiment, ALICE scours the debris from collisions between lead nuclei. Such impacts are thought to replicate the conditions present in the universe just after the Big Bang. Similar conditions might also be found at the heart of collapsing neutron stars.

LHCb The "b" stands for beauty, one of the six types of quark. Its main job is to look at why there is a mismatch between matter and antimatter in the universe. It has already found evidence of this "CP violation" in particles called D mesons.

ATLAS An acronym for A Toroidal LHC Apparatus, it is helping to look for evidence of theories that go beyond the Standard Model. Along with CMS, it was involved in the long-awaited discovery of the Higgs boson in 2012.

CMS The Compact Muon Solenoid is looking for evidence of other dimensions, as well as for evidence of supersymmetry. As one of the two detectors that found the Higgs, it will also play a big role in examining the new particle's properties.

39

Daily time discrepancy in GPS satellites unless relativity is corrected for (microseconds)

We live in remarkable times. The pre-space age era is still part of living memory for some—a time when humankind was limited to what it could achieve within the confines of the atmosphere. Today, over 1,000 satellites swarm around our planet. Thirty-two of them belong to the Global Positioning System, allowing you to locate yourself on the surface of the Earth with very high precision. GPS is run by the US military and became fully operational in 1995. The European Union hopes to have its version of GPS—called Galileo—fully operational by the end of the 2010s. India and China are planning alternatives, too. Thanks to these systems, all you need to do to find out where you are is whip out your smartphone, call up the maps function and a little dot will show you your location to within a few meters. And yet, if relativity wasn't taken into account, the whole system would be rendered useless within a day.

It's all relative

As their names suggest, Einstein's special and general theories of relativity state that everything is relative. In particular, your experience of time depends on your frame of reference. Time ticks at different rates in different situations. The first way you can alter your time (relative to someone else's) is to travel faster than them. Cosmonaut Sergei Krikalev is 0.02 seconds younger than he otherwise would have been thanks to his jaunts on Mir and the

International Space Station (see page 52). In a similar fashion, the highly accurate atomic clocks orbiting aboard the GPS satellites tick at a slower rate relative to those on the ground. They lose 7 microseconds a day.

However, correcting them by this amount would still lead to them being wrong. General relativity needs incorporating, too. Not only are the satellites orbiting at speed, they also experience a weaker gravitational pull from the Earth (relative to us on the ground). General relativity says that clocks in a weaker gravitational field run fast compared to those positioned in one that is stronger.

This speeds up the GPS clocks by 46 microseconds each day, so overall the clocks run fast by 39 microseconds.

Your smartphone is able to tell where you are because it fires radio signals off at least three of the orbiting satellites. The quicker the signal comes back, the nearer you are to that satellite. By triangulating the signals from all three, it can pinpoint your location. However, this relies on the clocks aboard the satellites staying in sync with the clock on your smartphone.

If the discrepancies caused by special and general relativity were not incorporated into the GPS network, the whole system would fall apart. Within a single day, the little dot on your smartphone would be up to 10 kilometers (6.2 miles) away from where you actually are. A mild inconvenience for the average user perhaps, but a major one for aviation or maritime traffic trying to safely navigate hectic routes.

▲ Artist's impression of a GPS-IIRM satellite in orbit. Without correcting for the effects of special and general relativity, the GPS system would quickly become useless.

43

Discrepancy in Mercury's orbit that Newton couldn't explain
(arc seconds/century)

Seen in books when we're young, the orbits of the planets look incredibly well behaved. They maintain their steady paths around the Sun, seemingly unaffected by each other's presence. Yet this picture is far too simple.

Johannes Kepler showed in the early 17th century that planets orbit not in circles, but in ellipses. This means a planet's distance from the Sun varies over the course of its orbit. Its closest approach to the Sun is called the point of perihelion. This point would be fixed if the Sun were the sole source of gravity in the system. However, the gravitational influence of the other planets causes this point to precess—it occurs at a slightly different point in the orbit each time. The resulting pattern of successive orbits is reminiscent of a spirograph.

Precession of the planets

This happens to all the planets in the solar system, including Earth. The exact amount of deviation is measured in arc seconds, where one arc second is $1/3600$ of a degree. Newton's theory of gravity could account for the precessing point of perihelion of all the planets, except one: Mercury. Newton's calculations suggested its perihelion should precess by 5,557 arc seconds percentury. Yet observing the planet shows the value is actually 5,600. The puzzle remained until the beginning of the 20th century.

When Albert Einstein published his general theory of relativity in 1915, he painted a different picture of gravity. He said that gravity is the result of the four-dimensional fabric of spacetime being warped by massive objects (see page 68). When general relativity is applied to the solar system, it churns out the correct amount of perihelion precession for all the planets. Where gravity is weaker, Newton's laws give a wonderful approximation of what's going on. However, the cracks begin to show in stronger gravitational fields—like the one experienced by Mercury, orbiting as it does so close to the Sun.

Despite this success, it was only a success on paper. For any scientific theory to be believed, it requires experimental evidence. That came four years later, in 1919. Astronomer Arthur Eddington traveled to Principe—a tiny island off the West African coast and one of the best places to view that year's total solar eclipse. During a solar eclipse, the Moon blocks out the light from the Sun, meaning stars that are normally washed out by sunlight are temporarily visible.

Newton and Einstein's theories disagreed about where these stars should be. Both physicists realized that the gravity of the Sun would cause the light from faraway stars to be bent slightly, leading us to see them in a slightly different position from where they are when the Sun isn't in the same part of the sky. However, Einstein said the light would be bent twice as much, as did Newton. When Eddington photographed the 1919 eclipse and measured how far the nearby stars deviated from their usual positions, he found in favor of Einstein. The mystery of Mercury's peculiar orbit was resolved and Einstein's theory of general relativity became the new explanation for the intricacies of gravity.

▼ A planet's closest approach to the Sun (perihelion) changes position over time. It took general relativity to accurately explain Mercury's precession.

Perihelion advances

Sun

Planet

45

Optimum angle to launch
a projectile (degrees)

Whether for sport or war, humans have been throwing projectiles for millennia. Cannonballs, arrows, bullets, shots, javelins, and basketballs are just some examples.

The path of a projectile is called its trajectory, and the total horizontal distance it travels before landing is known as the range. It is clearly an advantage in many situations to maximize the range of a projectile. The further you can throw your bomb, the further away from your target you are when it explodes; the further you can chuck your javelin, the more likely you are to pick up a gold medal.

Energy exchange

In practice, projectile motion is complex, with many factors involved. However, a simple picture can be imagined by ignoring any force other than gravity (i.e. air resistance). Imagine a tennis ball being thrown. It can be thrown at any angle between 0 degrees and 90 degrees relative to the ground. Clearly, throwing it at 90 degrees will result in a range equal to zero—the ball will go straight up, stop and then fall straight back down. During this flight there is an exchange of energy. The ball starts with some kinetic energy due to the speed at which it is thrown. As it climbs higher, this kinetic energy is converted into gravitational potential energy. The ball reaches the top of its climb once all kinetic energy is lost and is momentarily stationary. The ball then starts

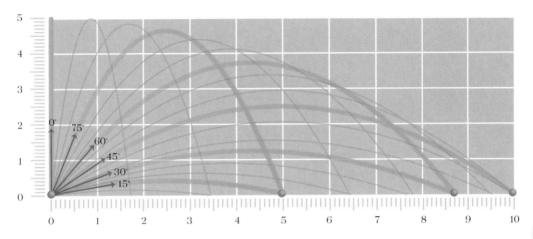

falling again as the gravitational potential energy is converted back into kinetic energy.

The ranges for projectiles launched at different angles. Forty-five degrees results in the object landing furthest away.

The shape of the flight path of any ball thrown at an angle less than 90 degrees will be a parabola. A key property of a parabola is that if it is cut in half at its highest point it is symmetrical—the path up is mirrored in the path down. This is because gravity only works to change the speed of a projectile in the vertical direction—its horizontal speed remains constant. So, to maximize the range of your projectile, you want the best compromise between giving it some initial sideways speed (which won't diminish) and giving it enough vertical speed to reach a good height so that it can gain enough gravitational potential energy for its downward trip. Doing the math shows that the best angle for achieving this is 45 degrees—exactly halfway between 0 and 90 degrees.

However, this is only true if the projectile is launched from the same height as it lands. For a football kicked from the ground or a long jumper launching from the runway this is fine, but it isn't for objects like shots and basketballs, which are thrown from approximately shoulder height. In these situations, the optimum angle tends to be a few degrees below 45. In many cases, air resistance cannot be ignored either, because it acts to slow the projectile down in the horizontal direction as well as the vertical. This also acts to lower the optimum angle for launch.

67.8

Hubble constant

(km/s/Mpc)

In 1929, US astronomer Edwin Hubble made a discovery that
shocked the world: the universe is expanding. Through his
telescope, he was able to see that not only were distant galaxies
moving away from us but also that the further away from us they
were, the faster they were moving.

Rising dough

To understand why this means the universe is expanding, imagine
some raisin bread dough put into the oven to bake. Let's say that
the raisins are initially placed at 1 centimeter (³⁄₈in) intervals and
overall the dough will double in size as it rises. Imagine what one
of the raisins would see. Any neighboring raisin that started off
1 centimeter away would be 2 centimeters (¹³⁄₁₆in) distant in the
finished bread (a change in distance of 1 centimeter). A raisin

▼ When dough rises, more
distant raisins appear to
move away faster. The same
is true of galaxies in an
expanding universe.

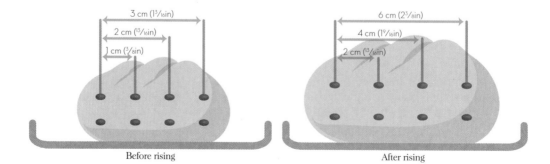

3 cm (1³⁄₁₆in)

2 cm (¹³⁄₁₆in)

1 cm (³⁄₈in)

Before rising

6 cm (2³⁄₈in)

4 cm (1⁹⁄₁₆in)

2 cm (¹³⁄₁₆in)

After rising

originally 2 centimeters ($^{13}/_{16}$in) away would end up 4 centimeters ($1^9/_{16}$in) away (a change in distance of 2 centimeters). Finally, a raisin initially 3 centimeters ($1^3/_{16}$in) away would eventually be 6 centimeters ($2^3/_8$in) away (a change in distance of 3 centimeters). If the bread took an hour to bake, the nearest raisins would have seemed to move 1 cm/h ($^3/_8$in/h); the furthest at 3 cm/h ($1^3/_{16}$in/h). So, despite the bread expanding at a constant rate, the raisin saw more distant neighbors moving away from it more quickly. That's exactly what Hubble found in 1929—the further the galaxy, the faster it appears to be receding. Like the dough, the universe must be expanding.

Hubble's constant tells us the rate of this expansion. In the raisin bread, the recession speed was 1 cm/h faster for every additional centimeter of original separation. The gap between galaxies is too big to be measured in centimeters. Instead, astronomers use a unit called the megaparsec, which is equal to 3.09×10^{22} meters. The galaxies are moving fast, too, so km/s is used instead of cm/h. Yet the principle is exactly the same. The most accurate value we have for Hubble's constant comes from the European Space Agency's Planck telescope, which measured it at 67.8 km/s (42 m/s) per megaparsec. So, for every additional megaparsec further away from us, galaxies are moving away 67.8 km/s faster.

This idea had profound consequences for cosmology. A universe expanding today must have been smaller yesterday. There must have been a time in the past when all the matter in the universe was concentrated in a single point. There must have been a beginning. Hubble's constant can be used to estimate when this origin occurred. This is done by dividing 1 by Hubble's constant and converting the megaparsecs into kilometers (that way, the units of kilometers neatly cancel, leaving you with an age in seconds). Using the Hubble constant measured by Planck, this is 4.55×10^{17} seconds or 14.4 billion years. In reality, this value has to be adjusted slightly to account for the shape of the universe. If this correction is applied to Planck data, the age of the universe comes out at 13.798 billion years (see page 132).

98
Number of naturally
occurring elements

The periodic table lists all known chemical elements in order of their atomic number—the number of protons they contain (which is what defines an element in the first place).

The modern periodic table currently contains 118 elements. The latest addition was ununseptium in 2010. However, it was artificially created by scientists in a laboratory, as were all the elements with atomic numbers 99 or above. Only the first 98 elements occur naturally.

It was once thought that only the first 92 elements occurred naturally on Earth but elements 93–98 (neptunium, plutonium, americium, curium, berkelium, and californium) have all been found in pitchblende—a radioactive, uranium-rich material. Of the 98, only 80 are stable—the other 18 are radioactive and so decay over time.

These elements are arranged within the periodic table in a series of rows and columns called periods and groups. In 1869, Russian chemist Dmitri Mendeleev published the first periodic table after realizing that the configuration of the elements periodically repeats. His table contained gaps that allowed him to predict the existence and chemical properties of eight new elements that had not been found at the time.

▲ Dmitri Mendeleev (1834-1907) published the first periodic table in 1869. The power of his work meant new elements could be predicted.

99.9999999999999

Percentage of empty space in a hydrogen atom

Despite making up everything you can see around you, atoms are incredibly barren places. In relative terms, the electrons orbit at an enormous distance from the nucleus. Sandwiched in between is an atomic no-man's-land of nothingness.

The approximate size of an atomic nucleus is 1 femtometer (10^{-15} meters), whereas the size of an entire atom is about 1 angstrom (10^{-10} meters). So, in terms of width, the atom is 100,000 times bigger than its central nucleus. However, in terms of volume, it is this number cubed. That makes the nucleus 1,000 trillion times smaller than the atom. Put another way, if an atom were blown up to the size of the Earth, the nucleus would be just 127 meters (416 feet) wide.

The fact that atoms are almost entirely empty space means that anything made of atoms is also mostly empty space. The book you're holding right now, the chair you might be sitting on, your eyes that are reading these words. It is quite remarkable to imagine something as solid as a chair actually being 99.9999999999999 percent empty. That space may be devoid of matter, but it isn't actually completely empty. It contains an electromagnetic field generated by the charged protons and electrons. Try to push the atoms in your body too close to the atoms in the chair and they will repel each other. As gravity is much weaker than the electromagnetic force, this repulsion prevents you falling through.

99.99999999874

Record speed in a particle accelerator (percentage of speed of light)

For all its fame and success, it isn't actually the Large Hadron Collider (LHC) that holds the record for the fastest speed ever achieved within a particle accelerator. That accolade goes to its predecessor, the Large Electron-Positron (LEP) collider, which was closed in 2000 to make way for the LHC. The LHC still uses the LEP tunnel, which was in operation from 1989. For comparison, the maximum speed attainable with the LHC is "just" 99.9999991 percent the speed of light. The LEP was able to reach a higher speed because it was accelerating much lighter particles.

One of the main aims of the LEP was to smash together particles in order to create and study W and Z bosons. This was important work as they had only been discovered at the beginning of the 1980s and their properties needed to be better understood.

What the speeds achieved during experiments like the LEP and the LHC demonstrate is that it is possible for humans to accelerate objects with mass to close to light speed. If, in the future, we could accelerate humans to similar speeds, it would open the door to significant time travel to the future, thanks to the rules of time dilation (see page 52). The particles in the LEP were experiencing time ticking over 200,000 times more slowly relative to our stationary clocks. If you were to travel at similar speeds on a round trip around space for what seemed to you to be a year, you would actually return to Earth 200,000 years into our future.

100

Boiling point of water (°C)

As you heat liquid water, its constituent molecules gain more and more energy. The water begins to boil when the thermal energy of the molecules exceeds the strength of the bonds between them. Once the water reaches 100°C (212°F), it will remain at this temperature until all of the liquid has boiled—any additional energy supplied at this point will go into breaking bonds rather than a further temperature rise. This period is known as a phase change. The latent heat of vaporization of a substance tells you how much energy you need to put in to turn 1 kilogram of a liquid into a vapor. For water this is 2,260 kilojoules.

The boiling point of water is used to calibrate the Celsius system devised in the 18th century by Swedish physicist Anders Celsius. On this scale, water boils at 100°C. At least, that is true of water at atmospheric pressure. The weight of the air pushing down on the water helps to keep the molecules close together, so more energy is required to liberate them. If you decrease the pressure, less energy is required to boil water. Under the right pressure, water can boil at room temperature. The boiling point of water can also be increased by adding materials like salt. This is called boiling point elevation.

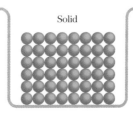

Solid

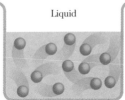

Liquid

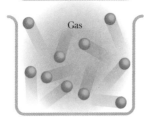

Gas

▲ Solids start off with their constituent atoms in close, regimented positions. As the material is heated, those atoms move further and further apart.

101.325

Standard atmospheric pressure (kPa)

The fact that you don't feel the weight of an entire planet's atmosphere pressing down on your shoulders is testament to how well evolution has sculpted our bodies to cope with life on Earth.

Measuring pressure

Atmospheric pressure is a measure of how much force is exerted on each square meter of the Earth's surface by the layers of air above it. At sea level, this is 101,325 newtons (1 pascal is one newton per square meter). This pressure can also be measured in atmospheres, with standard atmospheric pressure being 1 atm. The units of bars and millibars are often given as a measure of pressure, particularly on weather charts and in forecasts. A bar is equal to exactly 100,000 Pa—just a shade less than an atmosphere.

Your head and shoulders support an entire column of air and, as water does on the ocean floor, that air presses down on you. Let's say that your head and shoulders cover approximately 0.1 square meters. The weight of all the air molecules in the air column above you exerts a force upon you of around 10,000 newtons. That's literally a ton of air you are currently supporting. Luckily, our bodies can cope with this immense pressure, mainly because we have air inside us, which balances the pressure outside. The desire to maintain this balance between inner and outer pressure is exactly why your ears pop when you fly. Just be

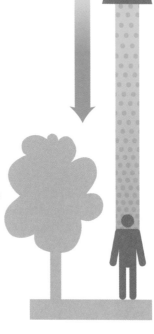

▲ The column of air immediately above your head contains a ton of molecules pressing down on you. Luckily we've evolved to cope.

Blaise Pascal (1623–1662)

French polymath Blaise Pascal, whose mother died when he was just three years old, was a child prodigy, writing a significant work on geometry aged just 16. He would later perform experiments with hydraulics, which led him to a deeper understanding of pressure. In doing this work, he also reportedly invented the syringe. Like many famous scientists, he has a crater on the Moon named after him.

He is also widely credited with inventing the world's first digital calculator—the Pascaline—perhaps inspired by his father's work as a tax collector. His father was later seriously injured to the extent that he was housebound. Two Jansenist priests looked after Blaise during this time, a period that is perhaps responsible for the devout religious faith he would develop later in life.

thankful you're not on Venus, where thick layers of carbon dioxide squeeze down to create an atmospheric pressure 90 times greater than Earth's. By contrast, Mars's tenuous atmosphere exerts a pressure at the surface just 0.5 percent of our planet's.

How air pressure changes with height for different planetary atmospheres can be calculated using the barometric formula. The change in pressure depends on several quantities and constants, including: the mass of an air molecule, acceleration due to gravity (see page 74), height, temperature, and Boltzmann's constant (see page 21). An alternative version can also be written in terms of the ideal gas constant (see page 73). Models of atmospheric pressure with height often assume that temperature drops by 2°C (35.6°F) for every 1,000 feet of altitude.

200

Terminal velocity of a skydiver (km/h)

Having been through all your training, you strap yourself into your seat just before the aeroplane accelerates down the runway and launches into the air. You see the dial on the altimeter rise: 5,000 feet, 10,000 feet, 12,000 feet. Jumpsuit on, you waddle to the open door at the rear. A few seconds later you are in a freefall. The sound of the air rushing past your head is almost deafening as your body tears through the atmosphere and accelerates toward the ground below.

That same air is acting on the rest of your body, too, and it is causing you to decelerate. Eventually, your acceleration drops to zero as the force of air resistance perfectly cancels the force of gravity pulling you down. A lack of acceleration doesn't mean you aren't still moving toward the ground, just that the speed with which you are doing so no longer increases. You have reached terminal velocity. For a skydiver who hasn't yet opened their parachute, this is around 200 kilometers (125 miles) per hour.

On October 14, 2012, Austrian skydiver Felix Baumgartner broke the record for the fastest ever freefall. Jumping from a height of around 39 kilometers (24 miles), he was able to reach a speed estimated to exceed 1,300 km/h (807 m/h). Starting from such a height, where the air is much thinner, allowed him to attain speeds far in excess of a conventional skydiver. On October 24, 2014, his record was broken by Google executive Alan Eustace, who successfully jumped from over 41 kilometers (25 miles).

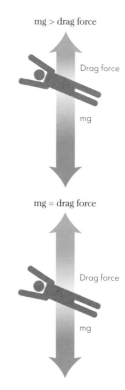

▼ Terminal velocity is reached when the force of gravity on your body (your weight, or mass times acceleration) is exactly matched by air resistance.

$mg >$ drag force

Drag force

mg

$mg =$ drag force

Drag force

mg

238

Mass number of uranium

While there are several natural isotopes of uranium, over 99 percent takes the form of uranium-238, which comprises 146 neutrons and 92 protons. The mass number of a chemical element reflects the combined number of protons and neutrons. Uranium occurs naturally in the Earth's crust at a rate of two to four parts per million, making it as common as tin.

Originally discovered by German chemist Martin Heinrich Klaproth in 1789, and first successfully isolated by Frenchman Eugène-Melchior Péligot 52 years later, the element was named after the planet Uranus, which had been discovered in 1781 (see page 72). However, it took until 1896 for its radioactive properties, for which it is arguably most famous, to be revealed. Henri Becquerel would share the 1903 Nobel Prize in Physics with Pierre and Marie Curie for his work on such radioactivity.

Unlike its isotopic cousin, uranium-238 cannot be directly used as a nuclear fuel. However, it can be used in breeder reactors to produce plutonium-239, which can be used in both nuclear power and nuclear weapons. This close relationship between uranium-238 and plutonium-239 was put to devastating use in the atomic bomb that was dropped on Nagasaki on August 9, 1945.

For many years it was believed that uranium had the highest number of protons of any naturally occurring element. However, trace quantities of six additional elements (including plutonium) have since been uncovered (see page 98).

▼ The ore from which uranium is extracted. While it is often thought of as a rare element, it is as common as tin.

331

Speed of sound in air (m/s)

Unlike light, sound needs a medium in which to travel. As the membrane of a speaker vibrates, for example, it causes the molecules in the air to vibrate, which in turn set off their neighbors. This results in a compression wave. Once the wave reaches your ear, the vibrations are turned into electrical signals that your brain interprets as sound.

The speed of sound is a measure of how fast the vibrational energy is passed from particle to particle. As such, it depends on the temperature, because that affects how strongly the air particles interact. At 0°C (32°F), the speed of sound in dry air is 331 m/s (361 y/s). At 20°C (68°F), this increases to 343 m/s (375 y/s). Either way, this means that the speed of sound in air at room temperature is approximately 1 million times slower than the speed of light. This is why, during a storm, you always see the lightning before you hear the thunder. The bigger the delay between lightning and thunder, the further away the storm is from you. A three-second delay is roughly equivalent to 1 kilometer (as 343 x 3 = 1,029).

Sound travels much faster in water than it does in air because the water molecules are huddled much closer together. At 25°C (77°F), the speed of sound in water is around 1,500 m/s (1,640 y/s)—over four times faster than in air at the same temperature.

▼ You see a lightning bolt before you hear a thunderclap because light travels faster than sound. The bigger the delay, the further away the storm.

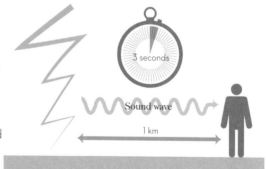

3 seconds

Sound wave

1 km

1,000

Density of liquid water (kg/m³)

As a liquid is cooled, its constituent molecules lose energy and start to move closer and closer together. At 80°C (176°F), for example, water has a density of 972 kg/m³. By the time it reaches 40°C (104°F), the atoms have moved closer together and the density rises to 992 kg/m³. So, it might be reasonable to assume that water is at its densest when transitioning to ice at 0°C (32°F). However, if that were true, ice wouldn't float in water. Our drinks wouldn't be as cool and the *Titanic* would never have struck an iceberg.

Water has a remarkable property unlike most other materials: it actually becomes less dense as it freezes. It is most packed together at 4°C (39°F), when its density is 999.9999985 kg/m³. This is crucial for life on Earth. Picture a pond in winter. Once the temperature of some of the water drops below 4°C, its density decreases, allowing the warmer, denser water to sink beneath it. This means lakes and rivers freeze from the top down, not the bottom up, allowing life to continue thriving under the ice.

As water turns to ice, the molecules arrange themselves into an ordered lattice pattern, rather than being randomly arranged as they are in a liquid. The gap between each point on the lattice is quite wide, so the molecules actually freeze into place further apart than they were in the liquid, leading to a lower density.

▲ As ice is less dense than water, an unusual property in a solid and its corresponding liquid, icebergs are able to float.

1,361

The solar constant (W/m^2)

This is the approximate amount of solar energy that falls on every square meter of the Earth. It may be called a constant, but in reality the amount of energy the Earth receives varies. One of the main reasons for this is that the Earth's orbit is not perfectly circular. The solar constant is defined as the amount of energy received at an exact distance of one astronomical unit—the *average* distance from the Sun to Earth (see page 140). The Earth receives more energy at its closest point (perihelion) and less at its further (aphelion). This leads to a 7 percent variation in energy received over the course of the year. There is a small, additional long-term variance of around 0.1 percent due to the fact the Sun goes through a cycle of varying magnetic activity, which peaks every 11 years or so. The first estimate of the solar constant was made in 1838 by French physicist Claude Pouillet. His answer of 1,228 W/m^2 is within 10 percent of today's accepted value, which is measured by low-flying satellites in orbit around the Earth.

The potential to power the planet

While it should be noted that this is the total energy received over all parts of the electromagnetic spectrum, not just light, there is still an enormous amount of energy arriving at our planet every second. The cross-section of the half of the Earth facing the Sun at any one time is around 127 million million square meters and every meter receives an amount of energy roughly equivalent to the solar constant every single second. If we could harness it, the amount of sunlight falling on us in just a single hour could power the entire planet's energy requirements for a whole year.

The Earth doesn't keep hold of all of the energy delivered by the Sun, however—it reflects some of it back into space. The fraction of the total incident solar energy reflected into space by a planet or other celestial body is known as its albedo. Different parts of the Earth reflect different amounts back into space. Sand has an albedo of 0.4; green grass comes in at 0.25. Ice is extremely reflective and has an albedo of around 0.9 (it reflects 90 percent of energy back into space). This might help contribute to periodic ice ages—as more ice forms, it reflects more energy into space, further lowering the temperature and creating more ice. The Earth currently has an average albedo of around 0.3, so we only keep hold of 70 percent of the energy we receive from the Sun.

From the amount of energy we receive, astronomers can calculate how much energy the Sun must be spitting out in order for us to receive that amount. The calculations suggest that we only catch about one two-billionths of the total energy the Sun spits out, making the total power output of the Sun just under 4 x 10^{26}W (see page 108).

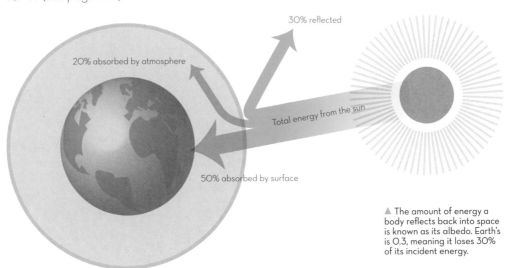

30% reflected

20% absorbed by atmosphere

Total energy from the sun

50% absorbed by surface

▲ The amount of energy a body reflects back into space is known as its albedo. Earth's is 0.3, meaning it loses 30% of its incident energy.

1543

Copernicus's *De revolutionibus* published

Perhaps it is no coincidence that the first recorded use of the word "revolution" in a political context—to overthrow one regime and replace it with another—appeared around 1600. Up until that point, it had mostly been used to describe the path of celestial bodies. Its new use may have been adopted in the wake of the events of 1543, when Polish mathematician Nicolaus Copernicus triggered a revolution of his own by overthrowing our ideas about the way these objects move.

His work *De revolutionibus orbium coelestium* ("On the revolutions of the heavenly spheres") is one of the most influential astronomical works ever published. In fact, such was its eventual impact that it is probably one of the most important books ever published on any subject. Yet its contents were mired in controversy. Legend has it that Copernicus only received the first printed copy on his deathbed.

Geocentric vs heliocentric

Copernicus's revolutionary idea was to place the Sun at the center of the solar system instead of the Earth. Ever since the days of the Ancient Greeks, the prevailing wisdom had us at the center of the universe. We were the solitary focus of attention for the Sun, planets and even the stars. This fitted well with the doctrine and dogma of the time. One of the biggest advocates for this so-called "geocentric" model was second-century polymath

▲ Polish mathematician Nicolaus Copernicus (1473–1543) controversially suggested that the planets orbited the Sun and not the Earth.

▼ Greek astronomer Ptolemy introduced the concept of epicycles—circular motions imposed on a greater circle—to explain the motion of the planets.

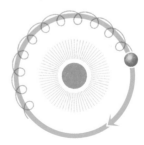

Claudius Ptolemy, but even Ptolemy realized things weren't as straightforward as this. Five other planets were known to the ancients (see page 72) and they were adept at tracking their movements across the sky. Such observations revealed a curious behavior: sometimes the planets seemed to halt their movement, before beginning to move in the opposite direction. This "retrograde" motion required an explanation. Keen to maintain the Earth's special place, Ptolemy introduced an ingenious trick: epicycles. An epicycle is a small circle on which a planet moves while simultaneously orbiting the Earth on a much bigger circle (called a deferent). A planet would seem to double back on itself during the part of the epicycle where it is moving in the opposite direction to its motion around the deferent. Such ingenuity saw Ptolemy's word remain unchallenged for well over 1,000 years. That is, until the publication of Copernicus's famous work.

By removing the Earth from its privileged position, and placing the Sun in the center, Copernicus could also explain the retrograde motion of the other planets. Take Mars as an example. Because we are closer to the Sun than Mars, we overtake it as we orbit. As we do, it seems to reverse its position relative to the background stars. Venus and Mercury appear to do a similar thing as they overtake us.

Due to its conflict with the ideas held dear by the Roman Catholic Church, Copernicus's revolution took a while to take hold. But Galileo's observations with the newly invented telescope in the early 1600s provided irrefutable evidence in favor of his heliocentric model and ushered in the era of modern astronomy.

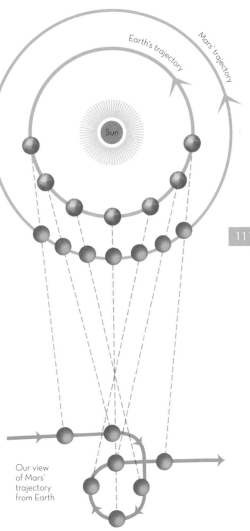

Earth's trajectory

Mars' trajectory

Sun

Our view of Mars' trajectory from Earth

▲ As Earth overtakes Mars in its orbit, the Red Planet seems to switch direction in our sky. This explanation replaced the earlier concept of epicycles.

1687

Newton's *Principia* published

Published in three parts, Isaac Newton's *Philosophiæ Naturalis Principia Mathematica* (Mathematical Principles of Natural Philosophy) first appeared on July 5, 1687. Contained within its pages are some of the most famous and important laws in all of physics. It not only contains his three laws of motion (see page 64), but also his Universal Law of Gravitation (see page 28). Its publication was arguably the biggest scientific revolution since Copernicus's 1543 work, and would stand unrivaled as physics' most important work until Einstein entered the fray in the 20th century.

The absence of any notes from his usually sizeable experimental notebooks in the years leading up to the publication of *Principia* shows just how absorbed Newton was in the production of his most famous work. His secretary would later recount that Newton was so consumed by it that he would sometimes forget to eat or sleep.

And yet it almost didn't get published. When Newton presented his manuscript to the Royal Society in London, they had already spent their publishing budget on a book called *The History of Fish* by Francis Willughby. British Astronomer Edmund Halley (he of comet fame) stepped in and personally paid for *Principia*'s publication. Halley's data on comets had helped Newton test his ideas on gravity.

In 2013, a first edition of Newton's book was sold at auction for nearly $500,000.

1,836.2

Ratio of proton and electron mass

Protons and electrons are two of the most important particles in nature, attracted to each other by their opposite electric charges. While the strength of their individual charges is identical (and opposite), they differ greatly in their masses. Due to the fact it is comprised of quarks, the proton is almost 2,000 times heavier than its atomic partner.

A big question is whether it has always been that way. Did this ratio take on a different value in the universe's younger days? Seemingly the answer is no. Astronomical observations suggest that the ratio has altered by a maximum of 0.00001 percent over the last 7 billion years. Astronomers looked at a galaxy that lies 7 billion years away, so the light we're seeing today carries with it information of what the galaxy was like when the light departed all that time ago. If the proton/electron mass ratio were different back then, the astronomers would see changes in the way that molecules of methanol—a type of alcohol—absorbed light. No such changes were seen.

Astronomers have probed even further back by looking at quasars—some of the most distant objects in the universe and therefore some of the oldest. Quasar observations indicate that the proton/electron mass ratio—often given the symbol μ—has changed by less than 0.001 percent in the last 12 billion years.

▲ While they share the same sized charge, protons are significantly heavier than electrons. This is because the former are made of quarks.

1905

Einstein's *annus mirabilis*

Few years have changed the face of physics like 1905. That year a 26-year-old clerk working in the Swiss Patent Office in Bern published four papers that would reverberate throughout the scientific world for decades. It was the beginning of Albert Einstein's journey to become the most celebrated and widely recognized scientist there has ever been. For this reason, 1905 is referred to as Einstein's "year of miracles." A century later, the United Nations would christen 2005 the "World Year of Physics" in commemoration. Here are the four papers he published in a remarkable 12 months.

▲ The year 1905 was a highly productive one for 26-year-old Albert Einstein (1879–1955). Arguably no year since has seen such revolutionary work.

ON A HEURISTIC VIEWPOINT CONCERNING THE PRODUCTION AND TRANSFORMATION OF LIGHT (published June 9)

This paper concerns the photoelectric effect. If a piece of metal is bombarded with radiation, the energy that radiation delivers can liberate some of the metal's electrons. However, certain things didn't make sense if, as widely believed at the time, light behaved solely as a wave. There was a minimum frequency required for the electrons to be ejected, for example. Also, once this threshold was reached, the electrons began escaping immediately. If light was behaving as a wave, it should take time for it to impart its energy.

Einstein's breakthrough was to suggest that the light must be being delivered in packets. So long as the packet had sufficient energy, it could liberate electrons immediately. He called these light packets "photons." He would win the 1921 Nobel Prize in Physics for this work.

ON THE MOTION OF SMALL PARTICLES SUSPENDED IN A STATIONARY LIQUID, AS REQUIRED BY THE MOLECULAR KINETIC THEORY OF HEAT

(published July 18)

This paper was on Brownian motion—the random motion of particles suspended in a liquid and first noted in 1827 by the botanist Robert Brown when observing pollen grains suspended in water. In his 1905 paper, Einstein explained this motion as the result of the pollen bouncing off individual water molecules and calculated the average amount by which they would be displaced over many such interactions.

ON THE ELECTRODYNAMICS OF MOVING BODIES

(published September 26)

Here Einstein introduces to the world what would become known as special relativity. In his paper he puts forward two postulates. One: the laws of physics are the same if you're not accelerating (i.e. you are stationary or moving at a constant speed). Physicists refer to such situations as an "inertial reference frame." Two: that the speed of light in a vacuum is constant. A consequence of such postulates bearing out is that time is relative. It directly leads to the idea of time dilation—that moving clocks run slow (see page 52).

DOES THE INERTIA OF A BODY DEPEND UPON ITS ENERGY CONTENT?

(published November 21)

Seemingly not content with his gargantuan efforts through the year so far, Einstein published his fourth paper. This time he argues that mass and energy are equivalent. Working through the mathematics, he shows that in some cases the exact relationship between mass and energy is $E=mc^2$.

5,778

Sun's surface temperature (K)

The Sun may be a giant seething ball of plasma, but its surface is actually slightly cooler than the center of the Earth, which scientists estimate to be around 500 K (440°F) hotter. Of course, being a ball of plasma, the Sun has no solid surface. Instead, the surface—called the photosphere—is the place in the Sun's atmosphere where photons are no longer confined by dense solar material and are free to wander out into space. At this point, the density is only about 0.37 percent of what it is in the Earth's atmosphere at sea level. The photosphere is thought to be around 100 kilometers (62 miles) thick (only 0.014 percent of the total solar radius).

If the Sun is treated as a perfect black body radiator (see page 36), then the Stefan-Boltzmann Law can be used to calculate the Sun's effective surface temperature, which comes out at 5,778 K (9,940°F). In reality, the temperature of the photosphere varies from 4,500 K to 6,000 K (7,640 to 10,340°F). Even cooler regions, called sunspots, appear from time to time. With

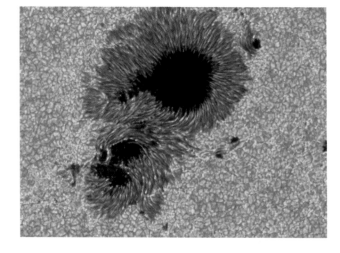

▼ Sunspots on the solar surface are areas that are cooler and darker—the result of magnetic activity inhibiting the flow of heat.

a lower temperature of 3,000 K to 4,500 K (4,940 to 7,640°F), they appear darker than surrounding regions. They are caused when magnetic activity inhibits the flow of energy from deeper layers within the Sun. Sunspots can be used to track the Sun's magnetic behavior, and 400 years of sunspot records show this activity rising and falling roughly every 11 years.

Not an average star

Not all stars in the universe have the same surface temperature. Normal stars are classified into seven main groups (labelled O, B, A, F, G, K, M in order of decreasing temperature). Each class is then subdivided into groups labelled O to 9 (again in order of decreasing temperature). The hottest stars have a surface temperature exceeding 30,000 K (53,540°F); the temperature of the coolest stars can drop as low as 2,400 K (3,860°F). A star's surface temperature is also responsible for its color—the hottest stars are blue and the coolest stars are red. The Sun, with its surface temperature of 5,778 K, sits somewhere in between and is a yellow G2 star. The vast majority (about 89 percent) of stars in the universe are cooler, redder K and M stars. Only around 7.5 percent are G stars like the Sun, so the often quoted fact that the Sun is just an "average" star is erroneous.

M K G F A   B   O

▲ Stars are classified by their color and surface temperature. The Sun is a G-class star; most stars are K or M.

Whether life can survive around K and M stars is still hotly debated among astronomers and astrobiologists. In order to maintain liquid water, a planet would have to huddle much closer to a red star's cooler surface. Doing so brings it within the "tidal lock radius"—the distance within which gravitational interactions between star and planet cause the latter to always present the same face to its host (much like the Moon does to the Earth). The dayside of the planet would be bathed in perpetual light, whereas the nightside would be plunged into constant darkness. How the climate evolves on such a world, and whether it is suitable for life, is a big open question in exoplanetary studies.

6,371

Radius of the Earth (km)

Our planet is not a perfect sphere: its rotation causes it to bulge at the equator, meaning the distance to the center of the Earth varies with latitude. The closest point on the Earth's surface to the center of the planet is found near the North Pole, at the bottom of the Arctic Ocean (6,353 kilometers, 3,947 miles). The furthest point on Earth from its core, at 6,384 kilometers (3,966 miles), is the peak of the Chimborazo volcano, near the equator in Ecuador. (Mount Everest is higher above sea level, but the volcano's proximity to the equatorial bulge means it is actually further from the center of the Earth.) The average radius of the Earth is 6,371 kilometers (3,958 miles).

Ancient Greek mathematician Eratosthenes was the first person to accurately estimate the Earth's circumference (and therefore its radius), in 240 BC. While there is still debate about exactly what value he came to, some say he got within 1.6 percent. Others think he was as much as 16 percent out. Either way, he achieved his estimate by measuring the elevation of the Sun at noon on the summer solstice from two different locations in Egypt. The difference in angle was equivalent to ¹/₅₀ of a circle. Knowing the distance between the two cities meant he could work out the length of the entire circle—the Earth's circumference. Indian mathematician Aryabhata (476-550 CE) later came up with an even more accurate value, which remained the best we had until modern times.

▲ Greek mathematician Eratosthenes (276-194 BC) was the first to accurately estimate the circumference of the Earth and hence its radius.

29,800

Speed of the Earth
around the Sun (m/s)

We're moving through space at quite a lick. In fact, the Earth covers a distance equivalent to its own diameter in just 7 minutes. Traveling at this speed, it takes our planet 1 year to complete a 940 million kilometer (585 million mile) lap of the Sun.

This is an average speed, however. Due to the Earth's orbit being elliptical, our planet speeds up when closest to the Sun (perihelion) and slows down when furthest away (aphelion). At perihelion, the Earth travels at approximately 48,280 km/s (30,000 m/s). However, Earth's orbit is very nearly circular, which can be used to simplify calculations and still give a good approximation.

There are many ways to estimate the speed of the Earth around the Sun. If you know the distance between the Earth and the Sun (see page 140) you can work out the circumference of its orbit if it were exactly circular (936 million kilometers, 581 million miles). It takes a year to cover that distance, so dividing distance by time would give 46,671 km/s (29,000 m/s).

You could also use Newton's second law of motion to work out the Earth's acceleration around the Sun due to our star's gravitational pull and get the speed from that. For this, you still need to know the distance between the Earth and the Sun, but you also need to know the mass of the Sun. Done this way, the answer comes out at 48,153 km/s (29,920 m/s). Both methods get you within 0.5 percent of the Earth's average orbital speed.

▼ The Earth's orbit around the Sun is not circular, but elliptical. This means our distance from the Sun varies.

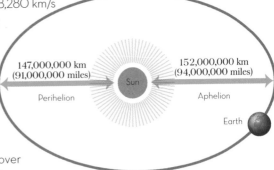

147,000,000 km
(91,000,000 miles)

Perihelion

Sun

152,000,000 km
(94,000,000 miles)

Aphelion

Earth

4,300,000

Number of lightning flashes on Earth per day

Our planet is literally crackling with electricity. An estimate dating back to 1925 suggested that a flashbulb of lightning goes off 100 times a second across the globe. Modern satellites have been able to keep a watchful eye on the planet and have yielded a more accurate value of 40–50. That's over 4 million lightning events per day. Clearly, such electrical discharge can be dangerous—estimates of the number of global fatalities due to lightning are often in the tens of thousands per year.

Unsurprisingly, lightning is most prevalent in sultry, tropical regions where about 70 percent of all flashes take place. The most lightning-afflicted location on Earth is the small, mountainous village of Kifuka in the Democratic Republic of the Congo, which experiences an estimated 158 lightning flashes annually.

Although the exact mechanism is still the subject of scientific debate, lightning occurs because of a difference in charge between the top (positively charged) and bottom (negatively charged) of a storm cloud. Rising moisture causes collisions that strip atoms of their electrons, which then gather at the bottom of the cloud. The now positively charged atoms continue to rise. If the electrons gather in sufficient number, they can repel the electrons on the Earth's surface, giving it an overall positive charge. Electricity can then follow from the bottom of the cloud to the ground.

▲ According to data from NASA, up to 50 lightning bolts occur around the world every second.

1.1×10^7

Rydberg constant (m^{-1})

Of all the fundamental constants in nature, this is the one that has been the most accurately measured—it has the least uncertainty in its value. It is also calculated from five of the other numbers in this book: the mass of the electron, the elementary charge, the permittivity of free space, Planck's constant and the speed of light (see pages 16, 22, 26, 12, and 126, respectively).

It is named after the Swedish physicist Johannes Rydberg, who published his work on atomic spectra in 1888. It later appeared in Niels Bohr's model of the atom (see page 27). According to Bohr, electrons can only orbit in a series of predetermined energy levels. When an electron drops from a higher energy level to a lower one, a photon of light is produced. Using the right equipment, this light can be seen as a series of bright bands known as spectral lines. Although he didn't appreciate what was causing them, Rydberg was looking at the spectral lines produced by alkali metals. He tried to produce a formula that would calculate where these lines should be. His sums included a constant, which now bears his name.

Combined with Bohr's ideas, it can be used to accurately predict the spectral emission from hydrogen atoms, no matter which energy level the electron starts and finishes in. The transition of an electron from the third energy level to the second, for example, yields a series of famous spectral lines known as the Balmer series.

▲ Swedish physicist Johannes Rydberg (1854-1919) lends his name to the most accurately measured constant in physics.

15 million

Core temperature of the Sun (K)

The surface of the Sun may be cooler than the center of the Earth, but the core of our star is much, much hotter. Around one-third of the Sun's substantial mass is squeezed into a region that equates to less than 1 percent of its volume, making the core more than a dozen times denser than lead. Under the crushing weight of the Sun's outer layers, the core pressure is at least 100 billion times greater than atmospheric pressure here on Earth. These extreme conditions, unrivaled in the solar system, raise the temperature to a staggering 15 million K.

Fueling fusion

In such conditions, two protons can overcome their natural electromagnetic repulsion and fuse to kick-start the proton-proton chain of nuclear fusion that powers the Sun (see page 48). Attempts to create fusion power on Earth require much higher temperatures, of around the 100 million K mark (see page 124).

The photons produced via fusion in the solar core have to wait a very long time before they make it out of the Sun. Such is the density of material in the Sun that, on average, the photons cannot make it more than a centimeter before bashing into something and being scattered off in a different direction. This intercollisional distance is known as the "mean free path." If the photons could continue outwards unhindered, they would make it to the edge of the photosphere in under three seconds. Instead,

▲ The work of Russian-American physicist George Gamow (1904–1968) was instrumental in deciphering how the Sun produces energy.

their meandering, bumper-car ride means they take at least 100,000 years to reach the inky blackness of space. They then take just over 8 minutes to make it all the way to Earth.

In order for two protons to fuse, they have to get within the range of the strong nuclear force, which can then overcome the electromagnetic force that would normally see like charges repel. That means they have to get within 10^{-15} meters of each other. However, the probability of two individual protons coming close enough together is one in a million trillion trillion (1 in 10^{30}). Suffice to say, that is extremely, extremely unlikely. However, the density of the core means there are also approximately 10^{32} protons per cubic centimeter. Just a few of those protons have the necessary speed to get close enough together to fuse. But, given the sheer size of the core, that's enough to maintain the Sun's energy output.

The role of fusion as the ringmaster of solar energy generation was first put forward by British astronomer Sir Arthur Eddington in 1920, just a year after he'd observed the eclipse that vindicated Einstein's general theory of relativity (see page 93). Eight years later, Russian-born physicist George Gamow published a paper detailing the fusion mechanism and included mention of what is now known as the Gamow factor—the probability that two protons will fuse together. According to classical physics, that probability is non-existent. However, when Gamow applied the new rules of quantum physics, he found that an effect known as "quantum tunneling" allowed for enough protons to fuse to keep the Sun shining.

Core

Radiation zone

Convection zone

▲ It takes an average of 100,000 years for a photon to navigate its way out of the Sun as it bounces between solar atoms.

16 million

World record for

fusion power (W)

It is well publicized that the human race may be heading for an energy catastrophe. Fossil fuels are running out, and their use so far has contributed to an increase in global temperatures. While renewable forms of energy, such as solar power, are already being deployed to meet some of our needs, it might not be enough to meet the demands of a growing population confronting a future without fossil fuels.

One possible solution is to take inspiration from the way the Sun creates power by turning mass into energy via fusion. Nuclear fusion is clean and green, and has the potential to create a wealth of energy. We cannot copy the Sun directly, however—we just can't generate the pressures and densities required to trigger the proton-proton chain. Instead of fusing protons together, fusion physicists combine deuterium and tritium (both isotopes of hydrogen). Both are relatively easy to get hold of, as deuterium can be obtained from seawater and tritium extracted from the lithium found naturally in the Earth's crust.

Hottest place in the solar system

The catch is that temperatures have to be even higher—100 million K—to get these heavier particles to fuse. When they do fuse, they create helium (an inert, non-polluting waste product) as well as a neutron. If the neutrons are allowed to collide with water surrounding the fusion machine, they can heat water to drive a turbine in a similar fashion to nuclear power plants. The real challenge is containing a stable 100 million K plasma. Such an environment is the hottest place in the solar system. It is done in machines called "tokamaks" which are, effectively, giant magnetic cages for encasing the superhot plasma. They get their name from a Russian word meaning "ring-shaped magnetic chamber."

▲ The Joint European Torus (JET) machine at the Culham Center for Fusion Energy, in England, holds the world record for fusion power.

The world record for the amount of fusion power produced was set by a team at the Joint European Torus (JET) experiment, housed at the Culham Center for Fusion Energy in Oxfordshire, England. When the machine was fired up, it managed to create 16MW of fusion power. The downside is that 24MW of power needed to be put in to achieve this. Putting more energy in than you get out is not going to work for a commercial power plant. Nevertheless, it proved that humans can imitate the Sun, creating energy by fusing subatomic particles together. The team at JET are busy working on ways to increase the efficiency of their tokamak by improving the way the giant magnets confine the plasma.

Currently, JET is the largest working tokamak in the world. However, construction is well under way on an even bigger fusion machine. It is hoped that the International Thermonuclear Experimental Reactor (ITER), based in France, will be capable of generating 500MW of fusion power—about the same as a small power plant. If successful, when it begins operation in around 2020, it will pave the way for the construction of a full-scale fusion power plant capable of meeting the energy requirements of an ever more technological future.

299,792,458

Speed of light in

a vacuum (m/s)

Given our everyday experience of light, it can be easy to imagine that it travels infinitely fast and doesn't take time to get from place to place. After all, we flick a switch and the light appears to come on instantly. Yet light does have a measurable speed, you just need astronomical distances for any resulting delay to become apparent. The insight into light's speed came as far back as 1676 with the work of Danish astronomer Ole Rømer, who was working at the Royal Observatory in Paris.

Eyeing up Io

Almost 70 years earlier, in 1610, Galileo had discovered the four biggest moons of Jupiter. Rømer was interested in one of these in particular: Io. Like all moons, Io orbits its planet in a very regular way, taking the same amount of time to complete each orbit. Yet Rømer noticed that sometimes Io was ahead or behind where it was supposed to be. These irregularities also depended on the time of year. He correctly reasoned that this must be because the light takes longer to reach us when Earth's orbit takes us further from Jupiter. If the speed of light were infinite, the differing distances wouldn't make a difference. The speed of light must be finite.

Rømer was able to estimate the speed of light from his observations, concluding that a beam of light takes around 11 minutes

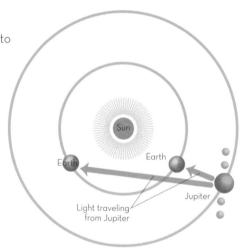

▲ The speed of light isn't infinite. When the Earth is further from Jupiter, we see the moons eclipse the planet at later times. The greater the distance, the more the delay.

Faster than light

In 2011, the physics world was brimming with the news that neutrinos had apparently been spotted moving faster than light. These supposed superluminal particles were dispatched from CERN, near Geneva, Switzerland, through the mountains of the Alps, to a detector in Gran Sasso, Italy. The length of their journey was around 731 kilometers (454 miles). According to the team behind the experiment, some of the neutrinos were arriving 60.7 nanoseconds quicker than a beam of light would if it had traversed the same distance.

There were three possible explanations. First, Einstein was wrong and the speed of light is not the fastest attainable speed in the universe. Second, the experiment was wrong and there was a fault somewhere that only made it appear as if the neutrinos were breaking the rules. The third possibility was perhaps the most tantalizing. The neutrinos weren't traveling faster than light, they just appeared to arrive ahead of schedule because they had taken a shortcut through another, unseen dimension. It would be a bit like someone setting a new record for the marathon by cutting a corner on the route. They wouldn't have run at record speed, but it would appear as if they had when they crossed the finish line.

It later transpired that the explanation was the least interesting one: a fault in the equipment. After painstaking analysis of the experimental setup, it was revealed that a fiber-optic cable had been incorrectly installed and a clock was ticking a little too fast. So, despite the brouhaha, it is still believed that the speed of light is the fastest you can travel through space.

to cover a distance equivalent to Earth's distance from the Sun. This would give a speed of light of around 220,000 km/s (137,000 m/s) in modern units. Today we know the answer is a shade below 300,000 km/s (186,000 m/s). The number is almost unimportant; it is the fact that Rømer realized that it takes time for light to get from A to B.

Unchanging speed

Not only is the speed of light in a vacuum finite, it is also fixed. This was discovered in 1887 as the result of the famous Michelson-Morley experiment. Imagine a moving walkway at an airport. If the walkway is moving at 1 m/s (3.3 ft/s), and you are walking in the direction of movement at 2 m/s (6.5ft/s), then an onlooker would see you moving at 3 m/s (9.8 ft/s, your speed plus the walkway speed). Conversely, if you tried walking against the walkway, the same onlooker would only see you moving at 1 m/s (your speed minus the walkway speed). In the 1880s there was no reason to think that light was any different. But Michelson and Morley's experiment showed that light doesn't play by the same rules.

Their elaborate equipment allowed them to effectively shine a light in the same direction as Earth's travel around the Sun. Six months later, they repeated the experiment when the Earth was moving in the opposite direction. As with the traveling walkway,

▼ The speed of light was expected to be slower if traveling against Earth's motion. It turns out the speed of light is constant.

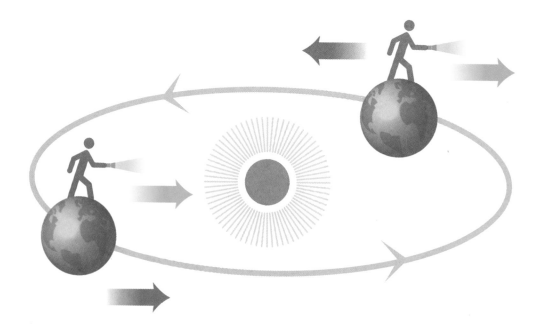

Resolving a paradox

Why is the night sky dark? This question is the essence of an old paradox that traditionally bears the name of German astronomer Heinrich Olbers, who wrote about it in 1823. However, the puzzle goes back to at least the 16th century.

If the universe is infinite in extent, with an infinite number of stars, then wherever you choose to look in the sky there should always be a star in your line of sight. This would cause the night sky to be bright and not dark. The fixed speed of light, along with the idea of an expanding universe, helps resolve the problem. There are parts of the universe being dragged away from us so quickly that light isn't able to get to us before it too is pulled away. So there are parts of the universe forever hidden from us because light from those regions is unable to reach us.

they expected to find light moving faster when moving with the Earth, and slower when moving against it. It turned out the Earth's motion was an irrelevance—the speed of light never wavered. Less than two decades later, Albert Einstein would use the fact that the speed of light is the same for all observers as one of his two cornerstone postulates of special relativity, which led directly to ideas about time dilation (see page 52). The same year would also see the first appearance of the equation $E=mc^2$, where c is the speed of light (although Einstein originally used the letter v for the speed of light, he switched to using c from 1907 onward). The "c" stands for "celeritas"—the Latin for swiftness.

1 billion

Density of a white dwarf
$(\mathrm{kg/m^3})$

A white dwarf is what's left behind when a small star puffs most of its envelope of gas out into space at the end of its life. This core contains about half of the mass of the overall star, and it is only about the size of the Earth. That much stuff, packed into a relatively small space, makes for an incredibly dense object. Every cubic meter of a white dwarf contains around 1 million tons of material. The density of the original star was only around 1 ton per cubic meter. For reference, osmium, the densest element in the periodic table, has a density of 22.6 tons per cubic meter.

While the density of a white dwarf sounds extreme, it is surpassed by other objects, both atomic and astronomical. A neutron star—the remnant of much more massive stars—has an average density 100 million times that of white dwarfs (see page 146). As does the atomic nucleus— the protons and neutrons are packed into such a small space that the density of the nucleus roughly equates to that of a neutron star. It is strange to think that your body and everything around you contains little pockets of density that exceed the density the Sun's core will reach when our star perishes in around 5 billion years' time.

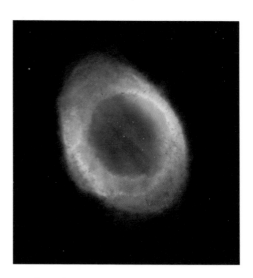

▲ The Ring Nebula (M57) is a planetary nebula—the fate that awaits the Sun. A white dwarf can be seen in the center.

9,192,631,770

Number of oscillations
of a cesium atom, which
defines a second

The Earth is a lousy timekeeper. The second used to be defined as $1/86,400$ of a mean solar day. However, the length of a day is not fixed. For starters, the Earth's rotation is slowing down due to gravitational interactions with the Moon (see page 42). Natural disasters can play their part, too—the earthquake that rocked Chile in February 2010 made the day 1.26 microseconds shorter. Seasonal changes in climate affect the strength of the winds blowing around the planet, in turn affecting how rapidly it spins.

In the modern world, where accurate timekeeping is crucial, relying on something malleable to define your base unit of time makes no sense. So physicists went searching for the most accurate replacement they could find. In the end, they turned to the element cesium. In 1967, the second was officially redefined to correspond to: "the duration of 9,192,631,770 periods of the radiation corresponding to the transition between the two hyperfine levels of the ground state of the cesium-133 atom."

It was later clarified that the cesium atoms had to be at rest at 0 K. Atomic clocks also have to be corrected to mean sea level due to gravitational time dilation (see page 52).

As the second and the speed of light are both defined to have exact values, they can be combined to define the meter. One meter is the distance covered by a beam of light in $1/299,792,458$ of a second.

▼ The world's first cesium atomic clock, unveiled in 1955 by the British National Physical Laboratory (NPL) in Teddington, Middlesex, England.

13.798 billion

Age of the universe (years)

Humans have long grappled with the age of the universe.
Remarkably, right up until the end of the 19th century, most people
thought the Earth was a few tens, or perhaps hundreds, of millions
of years old. Modern dating from radioactive rocks places the
true age of our planet at 4.54 *billion* years old.

Einstein wrong

Around the same time, the most widely held view of the wider
universe was that it was eternal—it had been around forever, so
asking its age was an irrelevance. In 1915, Einstein published his
general theory of relativity and was soon using it to construct
a model of the static universe. In the following few years others,
most notably Russian Alexander Friedmann and Belgian
priest Georges Lemaître, highlighted a flaw in Einstein's idea
and showed that the universe must be either expanding or
contracting. By 1929, American astronomer Edwin Hubble had
realized it was the former. And a universe expanding today must
have been smaller in the past. Suddenly an origin was on the
cards and asking the question of its age was back on the table.

From his work on galaxies, Hubble was able to find a value
for what has since been called the Hubble constant (see page
96). When converted into the right units, dividing one by that
constant gives an estimate for the age of the universe. Hubble's
original answer came out at 2 billion years. It was soon realized

Edwin Hubble (1889–1953)

Born in Marshfield, Missouri, astronomer Edwin Powell Hubble is one of the most recognizable names in the history of astronomy. The famous telescope that revolutionized the quality of space imagery in the 1990s was named after him.

At school and university he excelled most at sport, winning several gold medals on the athletics track and captaining the winning University of Chicago basketball team. He had promised his father that he would study law and became one of the first Rhodes Scholars when he spent three years studying the subject at the University of Oxford. Only after his father's death did he move into astronomy, receiving his PhD at the relatively late age of 28.

He arrived at the Mount Wilson Observatory two years later, just as the Hooker Telescope—then the biggest instrument in the world—was nearing completion. He would use the telescope to prove the existence of other galaxies outside our Milky Way, and discover that the further away they were, the faster they were receding. For the first time, it was known the universe might have a finite age.

this was unlikely to be correct and that the Earth was likely to be older. The first good estimate of the Hubble constant came from US astronomer Allan Sandage in the 1950s. Ever since, new generations of telescopes have gradually honed the value and hence the age of the universe. Recent results from the European Space Agency's Planck telescope have revealed the universe to be 13.798 billion years old. The previous value, from the Wilkinson Microwave Anisotropy Probe (WMAP), was closer to 13.7 billion.

93 billion

Diameter of the observable
universe (light years)

What's the furthest astronomers can see with their telescopes? The limit is not a symptom of technology—building a better instrument would not allow you to see any further. Instead, the restriction comes from the nature of the universe's birth. For the first 380,000 years of its existence, the universe was too hot to allow electrons and protons to combine into atoms. This melee of

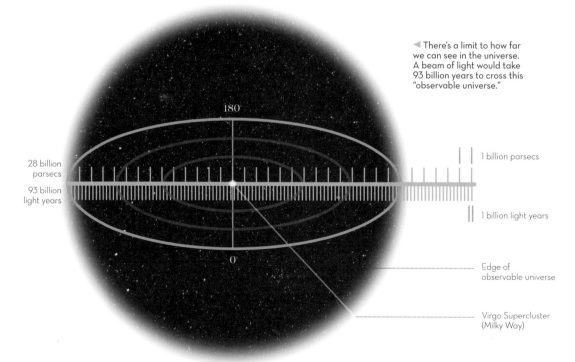

◀ There's a limit to how far we can see in the universe. A beam of light would take 93 billion years to cross this "observable universe."

28 billion parsecs

93 billion light years

180°

0°

1 billion parsecs

1 billion light years

Edge of observable universe

Virgo Supercluster (Milky Way)

atomic shrapnel kept light trapped, with photons bouncing around between protons and electrons and not being able to escape. Only when the temperature dropped enough to allow complete atoms to form could some of the light begin an uninterrupted journey toward the spot in which our planet would one day form to host astronomers with telescopes capable of capturing it. This earliest light forms the Cosmic Microwave Background (CMB, see page 58). It is the furthest we can see and forms the boundary of a sphere centered on the Earth, referred to as the observable universe—the bit of the whole universe we are able to see.

Expanding universe

Our best estimates suggest the Big Bang happened 13.8 billion years ago. So the CMB photons have been traveling across space at the speed of light for around 13.4 billion years— they appear to be coming from a spot 13.4 billion light years away. However, that is how far away from us that spot was when the light set off. While those photons have been heading our way, the universe has been expanding all the while, carrying that spot ever further from us. Models of the universe's expansion suggest that spot is now 46.5 billion light years away. That is the radius of our spherical, observable universe. Its diameter—the size of the observable universe—is therefore twice that figure, or 93 billion light years.

The center of the universe

It is worth noting that while we are, by definition, at the center of our observable universe, we are not at the center of *the* universe. That's because there is no center. At first glance it might seem like there should be. After all, the universe began with the Big Bang and has been expanding outward ever since. It may seem logical to think you can point to a place somewhere in the universe as the location at which the Big Bang happened. If a bomb went off in a room, for example, accident investigators could use the resulting shrapnel to determine where it detonated. The key difference between the two scenarios is that the room existed before the bomb went off. If the Big Bang were the bomb, the explosion also created the room. So if you point at any place in the modern universe and ask where that point was at the time of the Big Bang, the answer will always be in that first point of origin. So the Big Bang happened everywhere at once and there is no one special, central point.

100 billion

Temperature of a supernova (K)

When the supply of fusion energy created in a large star's core ceases, the star begins to buckle under the weight of its own considerable gravity. By this point the core is made of iron, so, as the surrounding layers rush toward the center they will soon encounter a very solid surface. The particles rebound from the core at almost the speed of light. As they rip through the higher, outer layers, the star explodes in a supernova with a temperature that can reach 100 billion K. A supernova is so bright that it often outshines all the stars in its host galaxy. Nearby examples, which provide a more detailed view, are very rare. Much of our thinking on this type of supernova comes from a 1987 explosion in the Large Magellanic Cloud—a dwarf galaxy that orbits our Milky Way. Known officially as SN 1987A, it was the closest supernova observed since 1604.

Creating new elements

Without supernovae, we wouldn't have nearly the breadth of chemical elements in the universe. Just after the Big Bang, the universe contained mostly hydrogen and helium with only trace quantities of lithium and beryllium. Over billions of years, stars turned some of that hydrogen and helium into heavier elements through fusion. However, stellar fusion ends at iron. The only way to continue making elements heavier than iron is for atoms to capture neutrons, becoming unstable isotopes. These isotopes can then undergo radioactive decay, which creates new elements. This mechanism of "supernova nucleosynthesis" was first put forward by British astronomer Fred Hoyle in 1954.

Neutron capture occurs in two main ways: the s-process and the r-process (slow and rapid). Slow capture can happen during the star's life and makes small amounts of elements heavier than iron. The r-process only works if the time it takes for the unstable isotope to decay is longer than the time it takes to capture a neutron. The heaviest element possible through this process is bismuth. The super-high temperature and the high density of neutrons during a supernova mean that the capture time is very short (hence the rapid). The unstable isotopes do not have time to decay before they are bolstered by more neutrons, bulking them up into heavier and heavier elements. Eventually the density drops and they can decay into new elements. Approximately half of all elements heavier than iron form through this r-process.

The force of the supernova explosion then flings these heavier particles out into interstellar space, where they are able to mix with other galactic material to form molecular clouds. Over time, these clouds collapse to form brand-new stars. A small amount of that material ends up in the planets that orbit the newly formed stars. So, without supernovae the Earth would have very few elements heavier than iron. Some of that material ends up in living things. As the famous American astronomer Carl Sagan once said: "We are made of starstuff."

▲ Hubble Space Telescope image of the Large Magellanic Cloud (top), and the location for the supernova explosion of 1987 (bottom).

125 billion

Approximate mass of the Higgs boson (eV/c^2)

The Standard Model of particle physics is one of the most successful theories physicists have (see page 82). Although unable to describe gravity, it explains the structure of the subatomic world and accounts for the behavior of the strong, weak, and electromagnetic forces. Its success is illustrated by its ability to predict the existence of particles later discovered through particle accelerator experiments.

Until recently, however, one major puzzle piece was missing. Why do some particles—like photons and gluons—have no mass, whereas particles like W and Z bosons are very heavy? In the 1960s, Briton Peter Higgs and Belgian François Englert independently provided an explanation, but they had to introduce an additional concept to the Standard Model, now known as the Higgs field.

All pervasive

Picture the Higgs field as a wide, expansive ocean permeating all of space. On Earth, animals can navigate their way through an ocean in different ways. A bird can fly right over, encountering no resistance from the water. A dolphin can surge forwards at great speed, jumping in and out of the water. An Olympic swimmer would be slower, but not as slow as a paddling dog.

In physics, a particle's mass is the result of how much it gets bogged down in the ocean of the Higgs field. Photons and

▲ Peter Higgs (b. 1929) and François Englert (b. 1932) were both awarded the 2013 Nobel Prize in Physics for their prediction of the Higgs boson.

gluons would be the bird—massless because they can move unhindered. Electrons are the equivalent of dolphins—only slightly restricted, resulting in a tiny mass. We humans would be quarks; the very slow-moving dog would be the considerably bulky W and Z bosons. Photons are the bosons responsible for the electromagnetic field, and the Higgs field needs a boson too— the Higgs boson. These are equivalent to the individual water molecules in the ocean analogy. The more a particle bumps into Higgs bosons, the more mass it acquires.

The idea was all well and good on paper, but the Higgs had to be found to be believed. Only then could it cement its place in the Standard Model. Not finding the Higgs would have been a considerable blow to the Standard Model's chances of being right. A particle that seemed to fit the bill was discovered at the Large Hadron Collider on July 4, 2012. By March 2013 it had been confirmed beyond reasonable doubt as the Higgs. Nine months later, Peter Higgs and François Englert were awarded the Nobel Prize in Physics.

Part of the decades-long delay in finding the Higgs was down to its own considerable mass. In order to produce particles in accelerators, you have to create collisions with enough energy to match the mass of what you are trying to find. This is because $E=mc^2$, which also dictates the units of mass that particle physicists use—eV/c^2 (eV is a unit of energy called the electronvolt, which is equal to 1.6×10^{-19} joules). The Higgs's mass came out at 125 billion electrovolts (125 GeV)—too heavy for previous accelerators to uncover.

▲ The Higgs Field is like an ocean that pervades all of space. Mass is the result of how much resistance you encounter from the "water."

50 billion

Distance between the Earth and Sun (m)

As we've encountered a couple of times, Earth's distance from the Sun changes day to day because our orbit is elliptical rather than circular. However, our average distance from the Sun is given by the astronomical unit (au). Since 2012, its exact value has been defined as 149,597,870,700 meters (490,806,660 feet).

Using such a value makes it much easier to get a sense of scale within our solar system;aying that Jupiter orbits the Sun at an average of 5.2 au (5.2 times further than we do) is far more convenient than giving its value as 778,547,200,000 meters (2,554,288,714 feet).

Cosmic distance ladder

But our distance from the Sun doesn't just tell us about our place in the solar system—it is a key tool in discovering our place in the wider universe. Astronomers use something called the "cosmic distance ladder" to work out how far away distant stars and galaxies are. Certain methods work well for nearby objects, but break down if you go too far away. These methods are like rungs on a ladder—you have to climb the first step before you can reach the second. The methods for nearby objects (the lower rungs) are used to calibrate the methods for faraway objects (the higher rungs). The very first rung on the ladder is the astronomical unit.

To see how this works, hold up a finger at arm's length in front of your face and close your left eye. Now line your finger up with

◀ James Cook (1728–1779, left) commanded a voyage sent to observe the Transit of Venus in 1769. The work of Johannes Kepler (1571-1630, right) was crucial in understanding how the planets orbit the Sun.

a distant object such as a tree, a door frame or the edge of a window. Quickly open your left eye and close your right. You'll see your finger move to the right with respect to the distant object. Get a sense of roughly how far it moved. Now repeat the exercise, but with your finger starting much closer to your face. You should see that your finger now jumps much further to the right. When viewed from two separate vantage points (i.e. the positions of each of your eyes), nearer objects move further than more distant ones. This is called parallax, and exactly the same trick can be used for nearby stars. Instead of using your eyes as the two observation points, a star is observed when the Earth is on one side of the Sun and then observed again six months later when the Earth is exactly on the opposite side. If you go through the geometry, you can work out how far away the star is from how far it has jumped compared to more distant, background stars. You just need to know the distance between your vantage points, which is twice the Sun-Earth distance (2 au). Without knowledge of the astronomical unit, we wouldn't know how far away the stars are.

While the value of the Sun-Earth distance put it at the top of the astronomical hit list, calculating it was exceptionally difficult. The ratio of the distances of the planets from the Sun had been known since Kepler published his laws of planetary motion in the early 17th century. Venus, for example, orbits the Sun at 0.72 au. If astronomers could somehow work out how far Venus was from the Sun in absolute terms, the astronomical unit could be easily calculated. That was easier said than done.

Transit of Venus

The only conceivable way to do it at the time was to use a fleetingly rare event called the Transit of Venus. This is where Venus appears to traverse the face of the Sun from our perspective, due to it being closer to our star. Parallax is again the key. This time Venus is the foreground object and the Sun the background object. If you can view the transit from

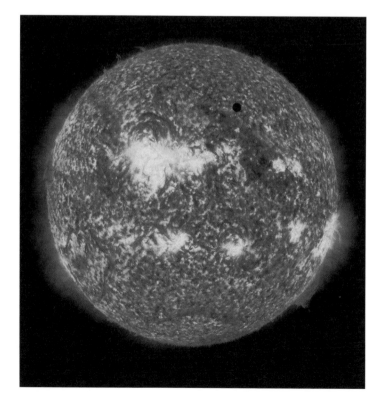

◄ Venus appears as a black dot on the upper right-hand side of the Sun's surface during the Transit of Venus in 2012.

two different places on Earth, the transit appears to begin and end at slightly different times. As long as you know the distance between your two locations, parallax can be used to turn this time discrepancy into the distance to Venus (and subsequently the astronomical unit).

Transits of Venus happen in pairs roughly eight years apart, but each pair is separated by more than a century. The last transit occurred in 2012, but the next won't be seen until 2117. Using his laws, Kepler became the first person to predict a Venus transit when he predicted the 1631 event in 1627. It wasn't visible in Europe, however, and so went unrecorded. Luckily it was the first of a pair and so astronomers didn't have to wait too long for the 1639 transit. On December 4 of that year, English astronomer Jeremiah Horrocks was one of only two people on the planet to observe and record the event. Horrocks used his observations to estimate the Earth-Sun distance as around 95,600,000 kilometers (59,400,000 miles, about two-thirds of what we know it to be today).

Such was the clamour for accurately knowing the astronomical unit—the key to unlocking the mystery of stellar distances—that by the transits of the 18th and 19th centuries, astronomers were dispatched all over the world to take measurements. The primary mission of James Cook's first voyage aboard HMS *Endeavour* was to observe the 1769 transit from Tahiti. In April 1770, the same expedition saw Cook and his crew become the first Europeans to set foot in Australia. French astronomer Jérôme Lalande used data from both the 1761 and 1769 transits to provide an updated estimate for the astronomical unit of 153 million kilometers (95 million miles, within around 2 percent of today's accepted value). Other astronomers obtained similar values. The transits of 1874 and 1882 refined the value yet further. Observations of the 2012 transit obtained a value within 0.007 percent. However, modern technology has seen new methods supersede parallax as the most accurate way of determining how far we are from the Sun. For instance, you can bounce radar off Venus and time how long it takes the signal to come back. Using methods like this, we can pin down the astronomical unit to within 30 meters (98 feet).

9×10^{13}

Mass-energy of 1 gram (J)

According to Einstein's famous equation $E=mc^2$, details of which he published in 1905, energy (E) and mass (m) are interchangeable (c is the speed of light). However, converting one into the other is an incredibly difficult process. The enormous, massive, and intensely hot Sun can only convert mass into energy through fusion with an efficiency of 0.007 (see page 48). And yet, if we could convert all of a large object's mass into pure energy, our concerns about energy supply would suddenly be laughable. Just one gram of matter would yield 9×10^{13} joules of energy.

It is estimated that the entire population of the planet consumed about 5×10^{20} joules worth of energy in 2010. In the same year, the energy locked up in our fossil fuel reserves was 100 times greater. Keep on consuming energy at that rate, and we'll work through it all in just 100 years.

What about converting mass into energy instead? How much mass is equivalent to our annual energy consumption? To work it out, all you need to do is divide that number by the speed of light squared. That comes out at about 5.5 tons (6 US tons)—less than the mass of an African elephant. Unfortunately we cannot convert mass into pure energy. Instead we are trying to power the future by copying the Sun and fusing subatomic particles to release meager amounts of energy (relative to what is theoretically available).

9.46×10^{15}

Distance traveled by
light in one year (m)

The vast distances in space require a new unit of measurement. One light year is the distance traveled by a beam of light in a vacuum in the course of one year according to the Julian calendar (exactly 365.25 days). As the speed of light is also defined to have an exact value (see page 126), this means the light year is exactly defined as 9,460,730,472,580,800 meters (31,039,141,968 feet).

In a similar manner, light seconds, light minutes and light hours can also be defined for objects within the solar system. The Moon is around 1.2 light seconds away, for example. The Sun sits just over 8 light minutes from us; on average Pluto is stranded 5.5 light hours from the Sun. Proxima Centauri, the nearest star to us (after the Sun), is 4.23 light years away; our Milky Way galaxy is 100,000 light years across.

Light years are often used in media aimed at the general public; however, an astronomer is more likely to use the parsec as a unit of distance. One parsec is equal to around 3.26 light years. Imagine a right-angled triangle with the Earth and Sun at each end of the shortest edge. If the angle opposite this edge is equal to one arc second ($^1/_{3600}$ of a degree) then the length of the triangle's base is equal to a parsec.

▼ A parsec is the length of the second-longest side of a right-angled triangle when the shortest side is 1 au and the smallest angle is 1 arc second.

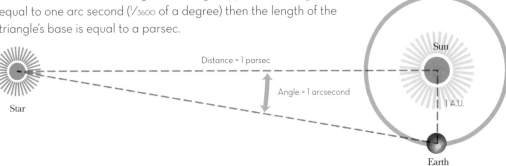

Distance = 1 parsec

Angle = 1 arcsecond

Sun

1 A.U.

Star

Earth

3.7×10^{17}

Density of a neutron star (kg/m^3)

▲ Dame Jocelyn Bell Burnell (b. 1943) and Antony Hewish (b. 1924) first discovered pulsars in 1967.

When a star dies, there are three options for what gets left behind. Small stars end up as white dwarfs (see page 130); gargantuan stars end with black holes (see page 162). The intermediate option is a neutron star. Such stars have cores whose masses exceed the Chandrasekhar limit (see page 56), and so cannot form a white dwarf.

As dense as an atom

As the core continues collapsing, the intense pressures force electrons to combine with protons to become neutrons. Like electrons in a white dwarf, these neutrons cannot continue to be compressed, otherwise it would violate the Pauli Exclusion Principle (see page 56). This "degeneracy pressure" props up the neutron star. However, neutrons can be pushed much closer together than electrons. A white dwarf is typically the size of the Earth, whereas a neutron star is often the size of a city (about 30 kilometers/18.6 miles across). Having a similar amount of material packed into a considerably smaller space sees the density skyrocket. At between 3.7×10^{17} and 5.9×10^{17} kg/m^3, neutron stars are the densest stars in the universe. Unsurprisingly, their density compares to that of the atomic nucleus where neutrons conventionally reside.

Such high densities can be hard to put into perspective. For example, one spoonful of material from a neutron star would weigh more than every person on Earth put together. A region of just 0.01 cubic kilometers would weigh as much as our entire planet. For your body to be that dense, all of your atoms would need to be squeezed into a space the size of a bacterium (a few millionths of a meter across). Suffice to say that a neutron star is a place of extremes.

Neutron stars also rotate exceptionally fast. In physics, a quantity called angular momentum has to be conserved. This is why the Moon is moving away from us as it slows down Earth's rotational speed (see page 42). It is also the reason an ice skater in a spin speeds up when they draw their arms in. Stars normally rotate once over the course of several weeks. A collapse to such a tiny size increases this speed, often to several spins a second. The fastest rotators—called millisecond pulsars—spin thousands of times a second.

▲ The Crab Nebula (M1) is perhaps the most famous example of a supernova remnant. It is thought to have exploded in 1054.

Maintaining magnetism

Another property that has to be conserved is related to magnetism. The strength of the star's magnetic field multiplied by its surface area must remain the same. As the surface area

The most famous pulsar

The most well-known pulsar is arguably the Crab Pulsar at the heart of the Crab Nebula in the constellation of Taurus. Back in 1054, Chinese astronomers recorded the appearance of a new star so bright it could even be seen during the day. We now know that this "star" was actually the explosive supernova that resulted in the Crab Pulsar. It rotates 30 times a second and is one of the most studied pulsars.

is dramatically reduced when it collapses into a neutron star, the strength of the magnetic field must shoot up to compensate. This high level of magnetism helps create beams of radiation which are emitted from each pole. If these beams are lined up with the Earth, we pick up pulses of radio waves as the neutron star rotates. Neutron stars are often referred to as pulsars (from "pulsating star") for this reason. The regularity of their pulses can lead to them keeping time more accurately than atomic clocks. The way two neutron stars in binary system interact is also an excellent test of Einstein's general theory of relativity.

▲ At the heart of the Crab Nebula lies a rapidly rotating neutron star—or pulsar—spinning 30 times a second.

Pulsars and Little Green Men

The first pulsar was heard on November 28, 1967, by Cambridge astronomers Jocelyn Bell and Antony Hewish. They could hear pulses separated by 1.33 seconds that couldn't be explained by any known earthly phenomenon. In reference to its regularity and extraterrestrial origin, the discovery was nicknamed LGM-1 (or Little Green Men-1). With subsequent discoveries came the idea that these were the rapidly rotating neutron stars that had first been predicted by Fritz Zwicky and Walter Baade in 1934.

Hewish would go on to win the 1974 Nobel Prize in Physics in recognition for his work on pulsars, becoming the first astronomer to do so. There has, however, been a great deal of controversy surrounding the fact that Bell was not also awarded the prize.

The link between pulsars and E.T. did continue. Both the *Voyager* and *Pioneer* space probes carry with them a map of the galaxy, showing the location of the Earth in relation to 14 pulsars.

6.02×10^{23}

Avogadro's constant (mol^{-1})

This is simply the number of molecules found in one mole of a substance and is most often given the symbol N_A.

Amedeo Avogadro was an Italian nobleman and scientist who built on French chemist Joseph Louis Gay-Lussac's work on gases. In 1811, Avogadro proposed that the volume of a gas at a fixed temperature and pressure depends on the number of molecules it contains. This principle would eventually be encapsulated in the Ideal Gas Law (see page 73). The value of Avogadro's constant was determined in the early 20th century by another Frenchman: physicist Jean Perrin. It was Perrin who suggested naming the constant after the Italian. He would win the 1926 Nobel Prize in Physics for his efforts to determine N_A. However, Austrian scientist Johann Josef Loschmidt made strides to calculate its value as early as the 1860s. In some texts—particularly those written in German—it is still referred to as Loschmidt's number.

One way to measure its value is to use the method of coulometry. This involves working out the electric charge on one mole of electrons and then dividing by the charge on a single electron (the elementary charge, see page 22). This gives the number of individual electrons that must be present. Today, X-rays can be used to probe crystals comprised of atoms to determine the value of N_A. Such is the precision of modern measurements that its value is known to eight decimal places.

▲ The number of molecules found in one mole of a substance is named after the Italian scientist Amedeo Avogadro (1776–1856).

2.2×10^{24}

Half-life of tellurium-128 (years)

A substance is radioactive if the nuclei in its consistent atoms are unstable and begin to spontaneously emit ionizing radiation (radiation with enough energy to strip electrons from atoms). This radiation comes in three varieties, named alpha (α), beta (β) and gamma (γ) after the first three letters of the Greek alphabet. In alpha decay, an alpha particle is emitted consisting of two

Natural radiation

We are regularly exposed to natural radioactive material. The biggest source is radon found in the air, released into the atmosphere by rocks. We breathe in this radon all the time and it has a relatively short half-life of just four days. The radiation produced when it decays can cause small amounts of damage to our lungs, which build up over time. It is estimated that radon is responsible for 1,100 lung cancer deaths each year in the UK. It is worth noting that half of these deaths occur in smokers. The average level of indoor radon in the UK is 20 Bq per cubic meter. This leads to a less than 1 in 200 chance of developing lung cancer for a non-smoker.

protons and two neutrons (the same as a helium-4 nucleus). Beta decay sees the emission of an electron (or positron). Finally, gamma radiation is simply a photon in the gamma ray part of the electromagnetic spectrum.

It is impossible to predict when any single atom will decay. However, it is possible to give a time after which half of the atoms in any given radioactive substance will have decayed. This is known as the half-life. Half-lives vary considerably for different materials. Hydrogen-7 (an isotope with one proton and six neutrons) has the shortest half-life at just 2.3×10^{-25} seconds. Tellurium-128 (52 protons, 76 neutrons) has the longest half-life at 2.2×10^{24} years. Therefore, it would take 100 trillion times longer than the current age of the universe for 50 percent of a radioactive sample of tellurium-128 to decay.

Famous faces

Radioactivity was first discovered in 1896 by Frenchman Henri Becquerel. The unit of radioactivity—the becquerel (Bq)—is now named after him. One Bq is equal to one atom decaying per second. An older unit for radioactivity is the curie (Ci), named after husband and wife duo Pierre and Marie Curie. One curie is equal to 3.7×10^{10} Bq—the radioactivity of 1 gram of radium-266, which the Curies worked on extensively. When Marie Curie died in 1934, the cause of death was linked to her exposure to radium. Pierre had died in 1906 after his skull was fractured when he fell under a horse and cart in Paris. All three scientists were awarded the 1903 Nobel Prize in Physics for their work, making Marie the first female recipient of a Nobel Prize. She would later become the first person to win two Nobel Prizes when she received the 1911 award for chemistry. To this day she remains the only woman to have been awarded twice.

▼ Henri Becquerel (1852–1908, top), Marie Curie (1867–1934), and Pierre Curie (1859–1906, bottom) were the founders of the field of radioactivity.

5.97×10^{24}

Mass of the Earth (kg)

How do you even begin to weigh a planet? The difficulty in doing so led to some interesting early ideas. Notable astronomers, including Edmund Halley, for example, believed the Earth to be hollow. In his 1692 work on the subject, Halley even suggested that gas escaping from our planet's vacuous interior was responsible for the Northern Lights.

A modern method for determining our mass was unavailable at that time despite the framework being laid by Isaac Newton's work on gravity, published in his *Principia* just a few years earlier, in 1687 (see page 112). Newton had derived Kepler's third law of planetary motion from his ideas on gravity. This law relates the mass of a body to the distance from, and orbital period of, any of its satellites. So, in theory, the Moon could have been used to weigh the Earth. The distance to the Moon had first been estimated by Greek astronomer Aristarchus in 240 BC; the Moon's orbital period is approximated by the duration of its cycle of phases.

Crucially, however, Newton's equations also contain his Universal Gravitational Constant, G (see page 28). An accurate value for G would not be obtained until 1798, by Henry Cavendish. If that information had been available earlier, a value of around 5×10^{24} kg would easily have convinced Halley to think again. Similarly, the mass of the Earth could have been determined from the acceleration due to gravity, g (see page 74), as Galileo calculated its value in the early 17th century. However, again, an accurate value of G was required to complete the calculation.

Nevil Maskelyne (1732–1811)

Despite being the first person to weigh the world, London-born Maskelyne is more famously associated with the quest to find longitude at sea. Sailors and their vessels were becoming increasingly marooned because they couldn't accurately determine their position east or west. In 1714, the British government passed the Longitude Act, which offered £20,000 to anyone who could solve the problem. The Royal Observatory at Greenwich had been founded in 1675 to help find an astronomical solution to the long-standing problem, and it was there Maskelyne found himself appointed Astronomer Royal in 1765.

The prize was famously won by clockmaker John Harrison, as detailed in Dava Sobel's book *Longitude*. Sobel cast Maskelyne as the villain to Harrison's heroics. However, there are some historians who disagree with this portrayal and there have been recent attempts to "rehabilitate" Maskelyne and his image in light of historical evidence. What's clear is that he was the first person to accurately estimate the mass of the Earth and all the planets then known in the solar system.

Weighing the world

The first evidence of Earth's considerable bulk would instead come from one of Halley's successors as Astronomer Royal—Nevil Maskelyne. Maskelyne took a team of scientists to Scotland in the summer of 1774 to conduct an experiment originally conceived by Newton himself. A mountain would be used to weigh the Earth. A pendulum hung next to a mountain experiences three forces: the force of the Earth's gravity downwards, the force of the mountain's gravity pulling it sideways, and the upwards force of tension in the string. The deflection by the mountain would cause the bob to move at a slight angle to the vertical. As the bob sits in equilibrium, all these forces must be balanced. The mass of the mountain and the mass of the Earth are responsible for the first two forces, so if the mass of the mountain could be obtained, and the amount of deflection in the pendulum measured, whatever was left must be the contribution from the mass of the Earth.

Maskelyne chose his mountain carefully, opting for Schiehallion (pronouced She-hal-e-on) in Perthshire. Far away from any other mountains, he could be confident the deflection was coming from Schiehallion alone. It is also almost perfectly conical, which allowed its volume to be determined relatively easily. Maskelyne's colleague, mathematician Charles Hutton, divided the mountain up into horizontal slices and worked out the volume of each slice before summing them. In doing so, he invented the concept of contour lines, which adorn every modern map. Mass is then density multiplied by volume, so as long as you can reliably estimate the density of the material making up the mountain, you can estimate the mountain's mass.

▲ Measuring the deflection of a pendulum toward the Schiehallion mountain allowed Nevil Maskelyne to calculate the mass of the Earth.

Clues about the core

While the experiment is pretty simple on paper, it was incredibly ardous in reality. Maskelyne and his men spent 17 weeks camped out on the slopes of Schiehallion, hampered by rain and desperately trying to obtain accurate measurements of the mountain and of the angle through which its gravity deflected the pendulum. Their efforts were ultimately rewarded with an estimate for the Earth's mass within 20 percent of today's accepted value. For the first time, it was clear that the Earth could not be hollow. Nor could it be rock all the way through—there must be an even denser material lying at depth. This was the first clue to Earth's bulky iron core, which generates our magnetic field and is therefore partly responsible for the Northern Lights. Despite his mechanism being totally wrong, Halley's link between the core and the shimmering curtains of light at the poles wasn't a complete red herring.

It seems that Maskelyne did miss a trick, however. Armed with the mass of the Earth, he could have produced an estimate for

the value of G, beating Cavendish to the feat by a couple of decades. However, that was never Maskelyne's goal. What he did acheive was to weigh not only the world, but also the entire solar system. At the time, without a value of G, the masses of the other planets were only known as fractions and multiples of the Earth's. Absolute values could now be placed on the other planets. Jupiter and Saturn were discovered to have a density less than rock, revealing them as gas planets, arguably for the first time. In modern times, Earth's mass, given the symbol M⊕, is used beyond the solar system too, allowing easy comparison of the mass of exoplanets to our own.

▲ The Scottish mountain of Schiehallion was selected by Nevil Maskelyne because of its conical shape, which simplified his calculations.

3.86×10^{26}

Energy output of the Sun per second (J)

The only way we can work out how much energy the Sun kicks out is to work backwards from how much we receive on Earth. Imagine a sphere around the Sun whose radius is equal to the distance between the Sun and the Earth. By working out how much energy falls on our little piece of the sphere's surface, we can use geometry to calculate how much energy the Sun must be sending through the entire sphere. This is the solar constant (see page 108) multiplied by the surface area of that sphere (which is 4π x Earth-Sun distance^2). This ends up as 3.83×10^{26} joules per second. Taking into account the fact that the Earth's orbit is elliptical and not circular raises the value slightly to the one stated at the top.

It is worth noting that this is the total energy output of photons of all wavelengths, not just visible light. The value is known as the Sun's luminosity and is often given the symbol $L_\odot$. Astronomers refer to luminosities where all wavelengths are considered as *bolometric* (as opposed to *visual* when only visible photons are included). $L_\odot$ is a convenient unit by which to compare the luminosity of other stars. The hottest, blue stars have luminosities of 10^6 (1 million) $L_\odot$; the dimmest, red stars have luminosities around $0.0001 \ L_\odot$.

1.29×10^{34}

Minimum half-life
of the proton (years)

For the last century and a half, many physicists have tried to unify the four fundamental forces (see page 66) into one coherent framework. This is often known as the Theory of Everything (TOE). According to the Standard Model (see page 82), protons are stable and do not decay. However, some attempts at TOE invoke proton decay in order to get their schemes to work. It would decay into a positron and a particle called a neutral pion.

Experiments have tried to spot evidence of proton decay, but as yet have been unable to do so. The Super-Kamiokande neutrino facility in Japan is one such experiment. A tank containing 50,000 tons (55,000 US tons) of water is buried 1 kilometer (0.6 miles) underground in an old mine. The tank is surrounded by over 11,000 light detectors which, among other things, are designed to pick up any flashes of energy produced by the positrons created by any protons decaying in the water. Since it began operation in 1996, no flashes related to proton decay have been observed.

However, by not finding anything so far, the experiment has placed a lower limit on the proton lifetime (if it has one). It is at least 1.29×10^{34} years. This has led to some versions of TOE being ruled out because they predict lifetimes that fall below this value. However, some other versions of TOE predict a proton decay half-life of around 10^{36} years, so proton decay might still be possible.

▼ While yet to be conclusively proved, the proton is thought to decay into smaller constituents over extremely long timescales.

Positron

Proton

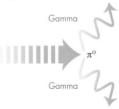

Gamma

π^0

Gamma

$\sim 10^{40}$

Ratio between electromagnetic and gravitational forces for a proton and electron

When it comes to the four fundamental forces of nature (see page 66), gravity really is the runt of the litter. Despite an entire planet pulling you down, it requires very little effort to jump a few inches off the ground or throw a ball high into the air. A magnet clings effortlessly to the fridge door.

Picture also the hydrogen atom, with its single electron orbiting a solitary proton. They feel an attraction to one another because of both their masses and their opposite charges. Yet over a distance equal to the Bohr radius (see page 27), their electromagnetic attraction trumps their mutual gravitational pull by a factor of 10,000 trillion trillion trillion (10^{40}).

Extra dimensions

Why gravity is so puny remains a mystery and the enigma is known as the Hierarchy Problem. One possible answer comes in the form of string theory. String theory says that quarks are made up of tiny strings that vibrate through more than the four dimensions of space and time that we experience in our everyday lives. These additional dimensions are curled up so small—on the scale of the Planck length—that we normally don't see them (see page 77). One explanation for gravity's relative weakness is that—unlike its fellow forces—it "leaks" through all the additional dimensions. Therefore, what we experience as gravity is a meek, watered-down version of its true might.

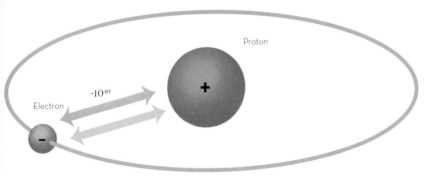

Proton

Electron

$\sim 10^{40}$

◄ Electrons and protons
feel two forces between
them: gravity and
electromagnetism. The
former is ~10^{40} times
weaker than the latter.

Having a quantum theory of gravity—a so-called Theory of
Everything—would help. Such a theory would describe the way in
which gravity was once united with the other three fundamental
forces before splitting out when the universe was around one unit
of Planck time old. In order to recreate these conditions, particle
accelerators like the Large Hadron Collider (LHC) would need
energies around the Planck energy—about 10^{18} times higher
than we can currently achieve. However, there may be a way in
through tackling a related puzzle.

We have not yet found a force-carrying boson for gravity. The
strong force has its gluon, the weak force its W and Z bosons
and the electromagnetic force its photon, but as yet the graviton
remains a hypothetical particle. Could it be so elusive because,
unlike the other bosons, it is likely to disappear into the other
seven dimensions before we have a chance to spot it? Physicists
working at the LHC are on the lookout for such behavior. If they
exist, it is possible that gravitons could be created by the particle
collisions inside the world's biggest atom smasher. If they then
skulk off into another dimension, they would leave behind a
telltale sign in data: a mismatch in the balance between energy
and momentum. These two quantities are always conserved
in particle collisions, so the size of the missing "gap" could tell
the physicists if the escapee is a close match to the theoretical
properties of the graviton. But, for now, the exact reason for
gravity's relative weakness is one of the biggest unanswered
question in modern physics.

$\sim 8 \times 10^{36}$

Mass of Milky Way's
central black hole (kg)

Look up on a clear night, far from the blinding lights of a city, and you'll see a dusty rainbow arching from one side of the sky to the other. This is our Milky Way galaxy. We see it like this because we are in it. Picture the Milky Way as similar to two fried eggs placed back to back. There is a thicker, almost circular yolk in the center, surrounded by a much flatter white. Astronomers call these two regions the bulge and the disc respectively. We live about halfway out from the center of the galaxy. So imagine what you'd see if you were inside the white of a fried egg—a white strip extending either

◄ A black hole is an area of space so warped by gravity that not even a beam of light can escape its grasp.

side of you and a thicker region in the center. That's very similar to how the Milky Way appears in our sky. Our Sun, along with all the other stars in the Milky Way, orbits around the center of the galaxy. It takes around 220 million years for the Sun to complete one galactic lap. So what are we all orbiting?

Invisible beast

Looking into the center of the Milky Way using telescopes sensitive to visible light is very difficult—our view is blocked by the immense amount of gas and dust between us and the center. Instead, astronomers use infrared telescopes to peer through the dust. One team, based at the University of California, Los Angeles (UCLA), has been studying the galactic center from Mauna Kea in Hawaii since the mid-1990s. Their observations show stars orbiting around a central point in just a few tens of years. Yet they appear to be orbiting an object we can't see in any form of electromagnetic radiation (light).

However, by employing Kepler's laws, the team can use the time it takes for the stars to orbit, coupled with their distance from the central point, to calculate the mass of the invisible object. Remarkably, it comes out at around 4.1 million times the mass of the Sun (or roughly 8×10^{36} kilograms). What's more, for those stars to remain in stable orbits, the central object must occupy a space much smaller than the solar system. The only object that fits the bill is a black hole (see page 162). That would also explain why it appears invisible to us—a black hole swallows everything, including all forms of light.

Yet our black hole is by no means the biggest. A galaxy called M87, which sits just over 50 million light years from us, has a central black hole that tips the scales at almost 7 billion solar masses (or around 1.4×10^{40} kilograms), so our black hole is tiny by comparison. M87 also has a huge jet of plasma that originates close to the central black hole and extends over 5,000 light years into space. It is thought supermassive black holes at the center of galaxies form when smaller black holes merge.

▲ The Milky Way arches over the Black Rock Desert in Nevada. A supermassive black hole lies at its center.

2×10^{67}

Decay time for a one-solar-mass black hole (years)

All stars die. Exactly how they do so depends on their mass. In their later years, stars like the Sun will switch from fusing hydrogen to fusing helium instead, but that is where the fusion ends. The gravitational crush of more massive stars can create core temperatures that exceed 100 million K. This allows them to keep on fusing heavier and heavier elements, steadily building up onion-like layers of elements such as carbon, oxygen, silicon, and sulphur. The process stops, however, when the result of fusion is iron—fusing iron into something heavier requires more energy input than it gives out, so it doesn't happen. What's left at the center of a massive dying star, then, is a super-dense iron core.

▲ The famous British physicist Stephen Hawking (b. 1942) has predicted that black holes evaporate over extraordinarily long timescales.

Inescapable

Now that fusion has ceased, there is nothing to prop the star up against gravity. Infalling material rebounds off the solid iron core and explodes outwards as a supernova. The iron core itself collapses into a vanishingly small point in a matter of seconds. All the mass of the original core is now concentrated into a point much smaller than an atom. Yet it still contains all that mass. The gravitational pull of such extremely dense material means that the escape velocity for this new object (see page 78) exceeds the speed of light. As the speed of light is the fastest attainable speed in the universe (see page 126), nothing should be able to escape. These stellar remnants are called black holes. The point at which the escape velocity exceeds the speed of light is known as the black hole's "event horizon."

Except, according to British physicist Stephen Hawking, the picture might not be that simple. It has long been noted that even a vacuum isn't completely empty (see page 23). Pairs of "virtual particles"—one particle and its corresponding antiparticle—pop into existence for a fleeting moment before annihilating back into energy. According to the rules of quantum mechanics—in particular the Heisenberg Uncertainty Principle—they must recombine to repay the energy debt they borrowed from the vacuum. But what if this process happens right on a black hole's event horizon? One particle would get sucked into the black hole, but the other would be free to escape. Forever separated, they can never recombine to pay off the energy debt. Hawking theorized that the energy must be repaid somehow and that the black hole itself emits a little energy to make up the shortfall. If true, this "Hawking radiation" means that a black hole isn't entirely black but glows ever so slightly.

This also means that black holes must slowly evaporate as they lose energy via Hawking radiation. This is a painstakingly slow process, however. The time it takes for a black hole to disappear depends entirely upon the black hole's mass. For a black hole with a mass equal to the Sun, it would take 2×10^{67} years— or more than 1 billion trillion trillion trillion trillion times the current age of the universe.

Event horizon

Hawking radiation

▲ If a particle/antiparticle pair is created on the event horizon of a black hole, it could lead to Hawking radiation and the black hole's eventual evaporation.

1×10^{80}

Estimate of the number of atoms in the visible universe

It is very difficult to know exactly how many atoms there are in the universe. However, estimates consistently place it somewhere between 1,078 and 1,082. One way to build up a good picture is to estimate how many atoms there are in a star, then how many stars there are in a galaxy, and finally how many galaxies there are in the visible universe.

Let's start with the Sun. It has mass of around 2×10^{30} kilograms and is mostly made of ionized hydrogen (a hydrogen atom with its electron removed, i.e. a proton). The mass of the proton is 1.6726×10^{-27} kilograms (see page 17), so the total number of atoms in the Sun is approximately 10^{57}. But the Sun is just one of between 200 and 400 billion (10^{11}) stars in our Milky Way galaxy. Taking the Sun as an average, the number of atoms in the Milky Way is then 10^{68}. In turn, the latest astronomical surveys suggest the number of galaxies in the observable universe is between 100 and 200 billion. So that makes the total number of atoms around 10^{79}. And then you've got all the atoms that you don't find in stars, such as those found in planets, moons, asteroids, comets and interstellar dust, and gas. This is tiny compared with the number of atoms found in stars, but probably adds one more power of ten—taking the total up to an estimated 10^{80} atoms.

5.2×10^{96}

Planck density $(\mathrm{kg/m^3})$

The expansion of the universe implies that the further back in time you go, the smaller it was. The logical conclusion is that there must have been a time when all the matter in the universe was squeezed into an incredibly small point. Unfortunately, we cannot get back to t=0 as our current rules of physics break down at this point. Instead, the earliest time we can talk about is the Planck time (see page 10). This is 5.39×10^{-44} seconds. Before this time, it is thought that gravity (which we describe using general relativity) and the other three fundamental forces (which we describe with quantum mechanics) were all united. At present, we have no coherent theory that unites general relativity and quantum mechanics, so we're restricted to describing the universe from the Planck time onward.

One question we can ask is how dense the universe was at this point. Like any density, this is mass divided by volume. Along with the Planck time and Planck length, the Planck mass and Planck volume can also be defined. When combined, they give a density of 5.16×10^{96} kg/m³. By using $E=mc^2$, we can also estimate the mass–energy density of the universe at this point. Multiplying the Planck density by the speed of light squared gives 4.63×10^{113} J/m³.

▼ The Big Bang model suggests the universe was born from a very small, very dense point just under 14 billion years ago.

1×10^{120}

Extent of the dark-energy vacuum catastrophe

In 1998, the astronomical world was turned upside down. Two teams of astronomers independently discovered that the expansion of the universe is speeding up—the exact opposite of what they expected.

Much as a star requires a balancing act between light pushing outward and gravity pulling inward, the behavior of the universe is governed by the balance between the force of its expansion and the mutual gravity of its constituent parts. Up until 1998, it was widely believed that the universe's expansion was slowing down as the force of its explosive birth petered out and gravity took an ever-increasing stranglehold.

Standard candles

Edwin Hubble originally discovered the expanding universe in the 1920s by noting that the more distant a galaxy, the faster it is receding from us (see page 132). He measured the distance to galaxies by using "Cepheid variables," which, as their name suggests, are stars whose brightness varies periodically. The time it takes for this pattern to repeat is related to the cepheid's intrinsic brightness. It will appear dimmer to us because the intensity of its light fades over distance. The further away the star is from us, the dimmer it seems. The discrepancy between intrinsic brightness and what we observe provides a measure of the distance. For this reason, cepheids are known as "standard candles." However, cepheids are no good if you want to look at more distant galaxies—i.e. further back into the universe's history—because they are simply too faint to pick out.

Fast forward to the 1990s, and the two sets of astronomers were using a brighter standard candle—type-1a supernovae—to explore the expansion rate of the universe earlier in its history by looking at more distant galaxies. These exploding stars have their origins in white dwarf stars. There is a restriction—called the Chandrasekhar limit—on the mass of a white dwarf (see page 56). If a white dwarf approaches this limit, either through ripping gas from a partner star or the collision between two white dwarfs, it explodes. As these stellar bombs always go off with a very similar mass, they should all have a very similar intrinsic brightness. Once again, we observe them as dimmer because of light fading over distance and the exact amount they fade is related to how far away they are.

The two teams repeated Hubble's experiment for more distant, older galaxies. If, as was believed, the universe's expansion is slowing down, those distant galaxies should be seen to be receding faster than those nearby. Exactly the opposite was found. Closer, younger galaxies—those whose light is telling us about more recent times—are the ones fleeing from one another at a greater rate. The rate of the universe's expansion is accelerating.

▲ If you know how bright an object should be, the dimmer it appears the further away it is. Such "standard candles" are widely used in astronomy.

Dark energy

What could be pushing galaxies ever further apart, against their natural desire to slow down under their mutual gravitational attraction? The answer is currently unclear, but this mysterious entity does have a name: dark energy. Observations suggest it makes up 68 percent of the mass–energy of the universe.

The leading contender for its true identity is called vacuum energy. Empty space is never truly empty (see pages 23 and 163). Even in a vacuum, pairs of virtual particles are popping in and out of existence all the time. According to quantum theory, this ever present energy must exert a repulsive force. What's more, unlike gravity, the strength of this force doesn't get weaker as galaxies move further apart. There must have come a time when the diminishing strength of gravity dipped below the fixed strength of the repulsive vacuum energy. At this point, galaxies would start to get pushed apart at an ever increasing pace.

There is a catch, however. When you compare the amount of energy locked up in empty space to the amount of energy required to explain the universe's accelerating expansion, the two don't match. They disagree by a factor of at least 1^{120}. This is arguably the biggest discrepancy between theory and experiment in the entire history of physics and is labelled the "vacuum catastrophe."

Alternative explanations have suggested that there is something wrong with the Chandrasekhar limit, so our calculations are giving us a false picture of what's happening to the universe. Others argue that gravity may work differently on really big, cosmological scales. So, despite the fact it makes up over two-thirds of our universe, we're still in the dark when it comes to dark energy.

1×10^{500}

Number of possible configurations in the string theory landscape

Some physicists consider string theory to be the best candidate for a Theory of Everything (TOE)—a single theory that is able to unite all four fundamental forces in nature (see page 66). Currently we use quantum physics to describe the strong, weak and electromagnetic forces and general relativity to explain gravity. Uniting those two regimes normally results in the mathematics falling apart. String theory has no such problems, at least on paper. According to string theorists, particles like quarks and electrons are not fundamental; instead they are made of tiny vibrating strings. They are curled up on the scale of the Planck length and vibrate across more dimensions than the four we normally encounter.

The main issue with string theory is that it is currently untestable—there are no experiments that can be performed to disprove or support it. String theorists talk of a "landscape" that contains all the different ways in which the strings can be configured. Estimates suggest this landscape contains 10^{500} possible configurations. Just one of them would represent the true nature of our reality. Some can easily be discarded as they don't give rise to the subatomic particles that we clearly know exist. But that smaller subset still contains a lot of possibilities and it is currently unknown which configuration our universe dances to. Without that knowledge, making testable predictions is very difficult.

▲ String theory says that quarks are made up of tiny strings that vibrate in different ways to yield different particles.

Further reading

Books

Aldersey-Williams, Hugh *Periodic Tales: The Curious Lives of the Elements* Penguin, 2012.

Al-Khalili, Jim *Quantum: A Guide for the Perplexed* Weidenfeld & Nicolson, 2003.

Birch, Hayley, Looi, Mun-Keat & Stuart, Colin *The Big Questions in Science: The Quest to Solve The Great Unknowns* Andre Deutsch, 2013.

Butterworth, Jon *Smashing Physics* Headline, 2014.

Carroll, Sean. *The Particle at the End of the Universe: The Hunt for the Higgs and the Discovery of a New World*. Oneworld Publications, 2013.

Chown, Marcus. *Quantum Theory Cannot Hurt You: Understanding the Mind-Blowing Building Blocks of the Universe*. Faber & Faber, 2014.

Chown, Marcus. *We Need to Talk About Kelvin: What everyday things tell us about the universe*. Faber & Faber, 2010.

Daintith, John, Gjertsen, Derek. *A Dictionary of Scientists*. Oxford University Press, 1999.

Ferguson, Kitty. *Measuring the Universe: The Historical Quest to Quantify Space*. Headline, 2000.

Feynman, Richard. *QED - The Strange Theory of Light and Matter*. Penguin, 1990.

Greene, Brian. *The Elegant Universe: Superstrings, Hidden Dimensions, and the Quest for the Ultimate Theory*. Vintage, 2000.

Gribbin, John. *In Search Of Schrodinger's Cat*. Black Swan, 1985.

Hawking, Stephen. *A Brief History of Time: From the Big Bang to Black Holes*. Bantam Press, 1988.

Jayawardhana, Ray. *The Neutrino Hunters: The Chase for the Ghost Particle and the Secrets of the Universe*. Oneworld Publications, 2014.

Krauss, Lawrence. *A Universe from Nothing*. Simon & Schuster, 2012.

Kumar, Manjit. *Quantum: Einstein, Bohr, and the Great Debate About the Nature of Reality*. Icon Books, 2009.

Liddle, Andrew. *An Introduction to Modern Cosmology (2nd Edition)*. Wiley-Blackwell, 2003.

Mahon, Basil. *The Man Who Changed Everything: The Life of James Clerk Maxwell*. John Wiley & Sons, 2004.

Miller, Arthur, I. *Empire of the Stars: Obsession, Friendship, and Betrayal in the Quest for Black Holes*. Houghton Mifflin Harcourt, 2005.

Oerter, Robert. *The Theory of Almost Everything: The Standard Model, the Unsung Triumph of Modern Physics*. Plume, 2006.

Orzel, Chad. *How to Teach Relativity to Your Dog*. Basic Books, 2012.

Panek, Richard. *The 4-Percent Universe: Dark Matter, Dark Energy, and the Race to Discover the Rest of Reality*. Oneworld Publications, 2012.

Sample, Ian. *Massive: The Hunt for the God Particle*. Virgin Books, 2011.

Smolin, Lee. *The Trouble with Physics: The Rise of String Theory, The Fall of a Science and, What Comes Next*. Penguin, 2008.

Sobel, Dava. *A More Perfect Heaven: How Copernicus Revolutionised the Cosmos*. Bloomsbury, 2012.

Waugh, Alexander. *Time: From Micro-seconds to the Millennia—a search for the right time*. Headline, 1999.

Young, Hugh & Freedman, Roger. *University Physics with Modern Physics (11th edition)*. Pearson Education, 2004.

Useful websites

"Official" String Theory www.superstringtheory.com

American Institute of Physics www.aip.org

CERN www.cern.ch

Einstein Papers Project www.einstein.caltech.edu

European Physical Society www.eps.org

Famous Physicists www.famousphysicists.org

Foundational Questions Institute www.fqxi.org

How Stuff Works www.howstuffworks.com

Hyperphysics, Georgia State University
http://hyperphysics.phy-astr.gsu.edu/hbase/hph.html

Institute of Physics www.iop.org

International Astronomical Union www.iau.org

International Bureau of Weights and Measures
www.bipm.org

International Center for Theoretical Physics www.ictp.it

International Union of Pure and Applied Physics
www.iupap.org

Minute Physics www.youtube.com/user/minutephysics

NASA www.nasa.gov

Physics arXiv blog www.medium.com/the-physics-arxiv-blog

Physics Central www.physicscentral.com

Physics Demonstrations www.physics.ncsu.edu/demoroom

Physics World www.physicsworld.com

PhysLink www.physlink.com

Splung www.splung.com

TED www.ted.com/topics/physics

The Nobel Prize in Physics
www.nobelprize.org/nobel_prizes/physics/laureates

The Particle Adventure www.particleadventure.org

The Physics Classroom www.physicsclassroom.com

Useful periodicals

American Scientist

Discover

Nature

New Scientist

Physics Today

Physics World

Popular Science

Science

Scientific American

Wired

Index

leptons 46
light see visible light
light years 71, 145
lightning 106, 120
Loschmidt, Johann Josef 149

M

M theory, dimensions in 76-7
Mars 34, 72, 79, 103
Maskawa, Toshihide 54
Maskelyne, Nevil 153-5
mass-energy of 1 gram 144
Maxwell, James Clerk 26, 39
Mendeleev, Dmitri 98
Mercury 72, 92-3, 111
mesons 46
Michell, John 29
Michelson-Morley experiment 128-9
Milky Way 145, 164
  central black hole 160-1
Millikan, Robert 22, 86
Minkowski, Hermann 68
Modified Newtonian Dynamics (MOND) 87
Moon 30, 42-3, 72, 74, 79, 145, 147, 152
muon neutrinos 63, 83
muons 53, 82

N

neutrinos 63
neutron, mass of 18
neutron stars 130, 146-8
Newton, Sir Isaac
  laws of motion 64, 93
  Principia 112
  Universal Law of Gravitation 28-9, 92, 152
night skies, darkness of 129
Novoselov, Konstantin 55

O

Olbers, Heinrich 129
Oort, Jan 87

P

pair production 35
parallax 141, 142-3
parsecs 145
Pascal, Blaise 103
Pauli Exclusion Principle 56-7, 146
Péligot, Eugène-Melchior 105
Penzias, Arno 59, 60
periodic table 98
Perrin, Jean 149
photons 13, 47
pi 65
Planck, Max 13, 85
  Planck density 165
  Planck length 11
  Planck time 10
  Planck's constant 12-13, 36, 65
Planck telescope 61, 133
planets known to ancients 72
Pluto 30, 72, 145
positrons 32, 82
Pouillet, Claude 108
projectile motion 94-5
protons 113
  ectromagnetic/gravitational forces in 158-9
  mass 17
  minimum half-life 157
  proton-proton chain 48-9, 122-3
Proxima Centauri 145
Ptolemy, Claudius 111
pulsars see neutron stars

Q

quanta 12
quantum chromodynamics (QCD) 54
quantum foam 11
quantum tunneling 123
quarks 17, 32, 46, 54
  see also specific types
quasars 50, 113

R

radioactivity 150-1
radon 150
redshift 38
rocket fuel, antimatter as 34
Rømer, Ole 126
Rubin, Vera 87
Rutherford, Ernest 24, 25, 27
Rydberg constant 121

S

Sagan, Carl 137
Sandage, Allan 133
Saturn 72, 155
Scott, Dave 74
Sommerfeld, Arnold 50
South Atlantic Anomaly 41
spacetime, dimension in 68-9
sparticles 89
specific heat 70
spectral lines 121
speed of light in vacuum 126-9
speed of sound 106
Standard Model 82-3
Stapp, Colonel John 75
stars
  distance to nearest 71
  neutron stars 130, 146-8
  surface temperatures 117
Stefan, Joseph 36
Stefan-Boltzmann constant 36, 116

Picture credits